AF588316

Lecture Notes in Chemistry

Volume 83

For further volumes:
http://www.springer.com/series/632

The Lecture Notes in Chemistry

The series Lecture Notes in Chemistry (LNC) reports new developments in chemistry and molecular science—quickly and informally, but with a high quality and the explicit aim to summarize and communicate current knowledge for teaching and training purposes. Books published in this series are conceived as bridging material between advanced graduate textbooks and the forefront of research. They will serve the following purposes:

- provide an accessible introduction to the field to postgraduate students and nonspecialist researchers from related areas,
- provide a source of advanced teaching material for specialized seminars, courses and schools, and
- be readily accessible in print and online.

The series covers all established fields of chemistry such as analytical chemistry, organic chemistry, inorganic chemistry, physical chemistry including electrochemistry, theoretical and computational chemistry, industrial chemistry, and catalysis. It is also a particularly suitable forum for volumes addressing the interfaces of chemistry with other disciplines, such as biology, medicine, physics, engineering, materials science including polymer and nanoscience, or earth and environmental science.

Both authored and edited volumes will be considered for publication. Edited volumes should however consist of a very limited number of contributions only. Proceedings will not be considered for LNC.

The year 2010 marks the relaunch of LNC.

Nicolas C. Polfer · Philippe Dugourd
Editors

Laser Photodissociation and Spectroscopy of Mass-separated Biomolecular Ions

Springer

Editors
Nicolas C. Polfer
Department of Chemistry
University of Florida
Gainesville, FL
USA

Philippe Dugourd
Institut Lumière Matière
CNRS - Université Lyon 1
Villeurbanne Cedex
France

ISSN 0342-4901 ISSN 2192-6603 (electronic)
ISBN 978-3-319-01251-3 ISBN 978-3-319-01252-0 (eBook)
DOI 10.1007/978-3-319-01252-0
Springer Cham Heidelberg New York Dordrecht London

Library of Congress Control Number: 2013946886

Printed on acid-free paper

Springer is part of Springer Science+Business Media (www.springer.com)

Foreword

The purpose of this book is to introduce students and researchers to exciting developments in laser-based photodissociation of trapped biomolecular ions inside mass spectrometers. The convergence of mass spectrometry (MS) and lasers in fact draws from a rich history in mass spectroscopy in the realm of physical chemistry and chemical physics. Lasers are also routinely employed to ionize biomolecules, for instance in matrix-assisted laser desorption/ionization (MALDI). The combination of MS with lasers allows researchers to obtain spectroscopic, and consequently structural information, on mass-separated ions. For example, based on infrared (IR) absorbances, the presence or absence of various chemical moieties can be verified. In addition, laser-based approaches often offer more control in ion activation. For instance, the absorption of an ultraviolet-visible (UV-vis) photon of a known energy at a known chromophore can trigger more selective photodissociation in the ion.

In this book, we will provide fundamental and operational background on the methods that are employed, and present on-going developments in applications in the context of biomolecules. In Chap. 1, a background on the vibrational and electronic structures of biomolecules is given, showing the influence of these factors on the corresponding IR and UV-vis spectroscopic signatures, as well as the photodissociation dynamics. Chapter 2 discusses benchtop tunable light sources that are useful for collecting IR and UV-vis spectra. In Chap. 3, the basics of different ion trap designs are discussed in terms of their usefulness for photodissociation experiments. Chapter 4 presents applications of vibrational spectroscopy on biomolecules. Chapter 5 delves into applications of UV-vis lasers for structural elucidation of biomolecular species.

Despite the tremendous potential for enhancing the structural information from MS measurements, laser-based biophysical and bioanalytical techniques have so far been limited to a handful of laboratories around the world. Two factors have contributed to holding back a broader implementation of these techniques: (1) researchers in the MS community often have limited familiarity with laser operation, and (2) there are technical challenges in coupling lasers to commercial mass spectrometers and carrying out these measurements. Some of these challenges are now being overcome. The on-going developments of turn-key benchtop light sources are extending these methods to nonspecialist groups. A new generation of mass spectrometers offers vastly higher sensitivity, thus providing the necessary signal-to-noise ratio to routinely carry out these measurements. It is the hope of the authors that this book will promote a wider understanding of the techniques that are involved, leading to an expansion of laser-based methods for the structural study of biomolecules by MS.

Acknowledgments

NP would like to thank the U.S. National Science Foundation for financial support under grant numbers CHE-0845450 and OISE-0730072. PD would like to thank the French National Agency for Research (ANR) and the European Research Council under the European Union's Seventh Framework Programme (FP7/2007-2013 Grant agreement N°320659) for financial support.

Contents

1 Spectroscopy and the Electromagnetic Spectrum 1
1.1 Light and Matter 1
1.2 The Nature of Light 2
1.3 Vibrations 3
1.3.1 Background 3
1.3.2 Biomolecular Vibrations 7
1.4 Electronic Transitions 9
1.4.1 Background 9
1.4.2 Applications to Biomolecular Systems 11
1.5 Absorption Spectroscopy 11
1.6 Consequence Spectroscopy 13
1.6.1 Background 13
1.6.2 Infrared Multiple-Photon Dissociation Spectroscopy 14
1.6.3 Ultraviolet Photodissociation Spectroscopy 16
1.6.4 Infrared–Ultraviolet Ion Dip Spectroscopy 18
1.6.5 Infrared Predissociation Spectroscopy 19
References 19

2 Light Sources 21
2.1 Laser Theory 21
2.1.1 Light Sources Employed in Photodissociation Experiments 21
2.1.2 Necessity of High-Intensity Light Sources 22
2.1.3 Amplification by Stimulated Emission 24
2.1.4 Population Inversion 26
2.1.5 3- and 4-Level Systems 27
2.1.6 Continuous Mode and Pulsed Mode (Q-Switching) 28
2.1.7 Cavities and Tuning 30
2.2 Gas Lasers 31
2.2.1 The CO_2 Gas Discharge Laser 31
2.2.2 Rotational States and Rovibrational Transitions of CO_2 33
2.2.3 Excimer Lasers and Other VUV Lasers 33

2.3 Solid-State Lasers 35
2.3.1 Nd:YAG 35
2.3.2 Widely Tunable Solid-State Lasers 37
2.4 Nonlinear Optics Light Sources 37
2.4.1 Optical Parametric Oscillators/Amplifiers 38
2.4.2 Nonlinear Crystals 39
2.5 Dye Lasers 40
2.5.1 Dye Laser Theory 40
2.5.2 Dyes Used 41
2.6 Semiconductor Lasers 42
2.6.1 Semiconductor Basics 42
2.6.2 Laser Diode Operation 43
2.6.3 Quantum Cascade Laser Operation 44
2.7 Other Light Sources 46
2.8 Concluding Remarks 47
References 47

3 Ion Traps 49
3.1 Introduction 49
3.2 Penning Traps 50
3.2.1 Basic Equations 50
3.2.2 Ion Excitation and Detection of the Signal 53
3.3 RF Traps 54
3.3.1 2D Quadrupole Ion Trap (Linear Ion Trap) 55
3.3.2 3D Quadrupole Ion Trap 60
3.3.3 Multipole Devices 65
3.4 Cold Traps 66
3.5 Comparison of Traps 67
References 70

4 Infrared Photodissociation of Biomolecular Ions 71
4.1 Infrared Photodissociation Spectra of Mass-separated Biomolecular Ions 71
4.2 Photodissociation with CO_2 Lasers 72
4.2.1 Background 72
4.2.2 Peptides 73
4.2.3 Saccharides 75
4.3 Photodissociation with Optical Parametric Oscillators 78
4.3.1 Background 78
4.3.2 Peptides 78
4.3.3 Saccharides 81
4.4 IR spectroscopy of Cold Ions 82
4.5 Longer-Term Considerations for IR Spectroscopy of Mass-separated Biomolecules 87

4.5.1 Current Constraints 87
4.5.2 Future Directions 88
References 90

5 UV-Visible Activation of Biomolecular Ions 93
5.1 Electronic Excitation of Biomolecular Ions 93
5.2 Gas-Phase Visible and Near-UV Excitation of Biomolecular Ions 94
5.3 UV Photodissociation of Protonated Peptides 97
5.3.1 Relaxation and Fragmentation of Electronic Excited Peptides: The Example of Leucine-Enkephalin 97
5.3.2 Peptide Sequencing with UVPD 99
5.4 Photoinduced Radical Chemistry 99
5.4.1 Background 99
5.4.2 UV Photodissociation of Phosphopeptides 101
5.4.3 Activated-Electron Photodetachment Dissociation 102
5.5 Coupling of UV-Visible Fragmentation with Separation Methods 104
5.5.1 Coupling of Photodissociation with Liquid Chromatography 104
5.5.2 Coupling of Photodissociation with Ion Mobility Spectrometry 105
5.6 Photodissociation of Biomolecular Ions with Ultrashort Pulses 107
5.6.1 Femtosecond Laser-Induced Ionization/Dissociation of Peptides 108
5.6.2 Femtosecond Pump-Probe Experiments on Trapped Flavin: Optical Control of Dissociation 108
5.7 Future Directions 109
References 116

Index 117

Chapter 1
Spectroscopy and the Electromagnetic Spectrum

Corey N. Stedwell and Nicolas C. Polfer

1.1 Light and Matter

Electromagnetic radiation exhibits properties of both particles and waves, known as the wave–particle duality. In the wave model, electromagnetic radiation is characterized by its *frequency*, v, *wavelength*, λ, and velocity, c. These three values are related by the relationship

$$c = v\lambda \tag{1.1}$$

The value of c is constant in a given medium (e.g., 2.99×10^8 ms^{-1} in vacuum), while the frequency and wavelength of light are inversely proportional to one another. The SI units for wavelength and frequency are the meter (m) and the hertz (Hz), respectively. Traditionally, spectroscopists also define electromagnetic radiation by the unit *wavenumbers*, defined as

$$\tilde{v} = \lambda^{-1} \tag{1.2}$$

where λ denotes the wavelength in centimeters.

The energy of a photon (quantum of electromagnetic radiation) depends solely on its frequency (or wavelength) and is defined as

$$E = hv = \frac{hc}{\lambda} = hc\tilde{v} \tag{1.3}$$

where h is Planck's constant (6.63×10^{-34} Js) [1]. Note that energy is directly proportional to frequency and wavenumber, and inversely proportional to

C. N. Stedwell (✉) · N. C. Polfer
Department of Chemistry, University of Florida, Gainesville, FL 32611, USA
e-mail: cstedwell@ufl.edu

N. C. Polfer
e-mail: polfer@chem.ufl.edu

N. C. Polfer and P. Dugourd (eds.), *Laser Photodissociation and Spectroscopy of Mass-separated Biomolecular Ions*, Lecture Notes in Chemistry 83,
DOI: 10.1007/978-3-319-01252-0_1,

Table 1.1 Summary of the frequency, wavenumber, and energy ranges for several regions of the electromagnetic spectrum

Region	Wavelength/m	Wavenumber/cm^{-1}	Energy/eV
Ultraviolet	4.0×10^{-7}–1.0×10^{-8}	2.5×10^{4}–1.0×10^{6}	3.1×10^{0}–1.2×10^{2}
Visible	8.3×10^{-7}–4.0×10^{-7}	1.2×10^{4}–2.5×10^{4}	1.5×10^{0}–3.1×10^{0}
Near-IR	1.5×10^{-6}–8.3×10^{-7}	6.7×10^{3}–1.2×10^{4}	8.3×10^{-1}–1.5×10^{0}
Mid-IR	1.0×10^{-5}–1.5×10^{-6}	1.0×10^{3}–6.7×10^{3}	1.2×10^{-1}–8.3×10^{-1}
Far-IR	1.0×10^{-4}–1.0×10^{-5}	1.0×10^{2}–1.0×10^{3}	1.2×10^{-2}–1.2×10^{-1}
Microwave	1.0×10^{-1}–1.0×10^{-4}	1.0×10^{-1}–1.0×10^{2}	1.2×10^{-5}–1.2×10^{-2}

wavelength. General ranges for frequency, wavelength, and energy limits for several regions of the electromagnetic spectrum are given in Table 1.1.

1.2 The Nature of Light

Visible (vis) light, microwaves, X-rays, and so forth are all different kinds of electromagnetic radiation. Taken collectively, they make up the *electromagnetic spectrum*, depicted in Fig. 1.1.

The electromagnetic spectrum is arbitrarily divided into regions, and it is clear that the familiar, vis portion of the spectrum (i.e., 390–750 nm) constitutes only a small section of the full spectrum. The vis region is flanked by the higher-energy, ultraviolet (UV) and the lower-energy, infrared (IR) and microwave regions. These frequency ranges are the most widely used in spectroscopic measurements, as they can each provide key structural insights into analyte molecules. The microwave

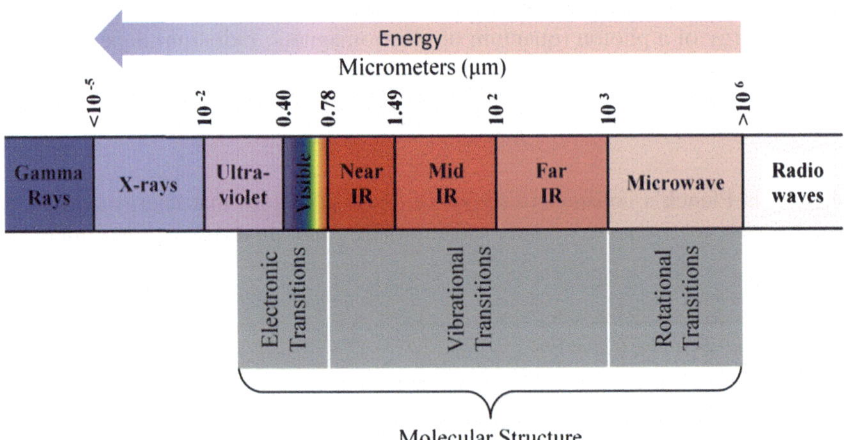

Fig. 1.1 The electromagnetic spectrum covers a continuous range of wavelengths, from low-energy radio waves to gamma (γ) rays at the high-energy end

region of the electromagnetic spectrum is associated with rotational spectroscopy, while the UV and vis portions correspond to molecular electronic transitions. The IR region, which is typically broken into near-, mid-, and far-IR, is utilized in vibrational spectroscopy, as the frequencies of radiation therein correspond to molecular vibrations.

1.3 Vibrations

1.3.1 Background

There are two general types of molecular vibrations, namely stretching and bending. The stretching frequency of a bond can be approximated by Hooke's law [2, 3]. In this approximation, two atoms and the connecting bond are treated as a simple *harmonic oscillator* composed of two masses (atoms) joined by a spring (a chemical bond). This is illustrated for atoms, labeled A and B, in Fig. 1.2.

These two atoms have some equilibrium distance, r_e, and as the atoms are displaced from their equilibrium distance, they will experience a restoring force, F, that opposes the motion. If we assume that the system behaves according to classical mechanics, the restoring force will be proportional to the displacement from equilibrium distance, Δr, and vary according to Hooke's law

$$F = -k\Delta r \tag{1.4}$$

where k is the *force constant*. At some point, the restoring force will cause the molecular motion to cease, and subsequently, the atoms will begin moving in the opposite direction. This creates a smooth oscillatory motion known as a vibration.

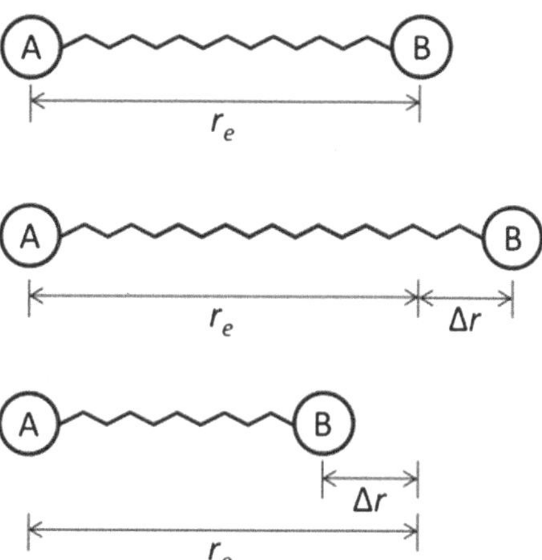

Fig. 1.2 Motions of a diatomic oscillator AB

The frequency of vibration of the system is related to the mass and the force constant of the spring, by

$$\nu = \frac{1}{2\pi}\sqrt{\frac{k}{\mu}} \tag{1.5}$$

where μ is the *reduced mass* and is defined as

$$\mu = \frac{m_A m_B}{m_A + m_B} \tag{1.6}$$

with m_A and m_B being the mass of atoms A and B, respectively.

In the classical harmonic oscillator, the energy of the vibration is given by

$$E = \frac{1}{2}k\Delta r^2 = h\nu \tag{1.7}$$

Thus, the energy (or frequency) is dependent on how far one stretches or compresses the spring. In a classical picture, any value of r is possible. If this model were true, a molecule could absorb energy of any wavelength. However, vibrational motion is quantized (it must follow the rules of quantum mechanics), and the only allowed transitions fit the equation

$$E(\mathbf{v}) = \left(\mathbf{v} + \frac{1}{2}\right)h\nu \tag{1.8}$$

where ν is the frequency of the vibration and $\mathbf{v}$ is the *vibrational quantum number* (0, 1, 2, 3, …). The concept of the quantum-mechanical harmonic oscillator potential energy curve is illustrated in Fig. 1.3, indicating the vibrational states.

Note that ν does not depend on the value of the quantum number $\mathbf{v}$. Hence, the molecular vibrational frequency is the same in all states, even though the energy E changes with $\mathbf{v}$ in Eq. (1.8). The lowest possible energy level, $\mathbf{v}_0$, resides above the bottom of the energy curve; this difference is known as the *zero-point energy*.

The *gross selection rule* for vibrational transitions stipulates that the electric dipole moment of the molecule must change during the course of a vibration. In other words, the so-called *transition dipole moment*, μ_{fi}, between states i and f must be non-zero. More rigorously,

$$\mu_{fi} = \int \Psi_f^* \hat{\mu} \Psi_i \neq 0 \tag{1.9}$$

where Ψ_i and Ψ_j are wavefunctions describing the ith and jth states, * denotes the complex conjugate of the wavefunction, and $\hat{\mu}$ is the electric dipole moment operator. If the value of this integral is non-zero, then the vibrational mode is said to be *IR active*, meaning that the transition is allowed and hence observed. A vibrational mode is said to be *IR inactive* if the dipole moment of the molecule does not change during the vibration, as for instance in a homonuclear diatomic molecular vibration.

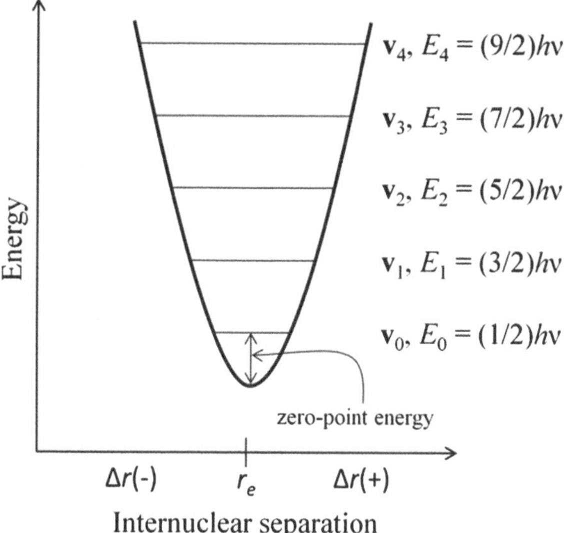

Fig. 1.3 Quantum-mechanical harmonic oscillator indicating allowable energy levels

The *specific selection rule* for the harmonic oscillator limits changes in the vibrational quantum number to $\Delta\mathbf{v} = \pm 1$. Substituting any state $\mathbf{v}$ and $\mathbf{v} + 1$ in Eq. (1.8), we can see that the energy difference between any two states is

$$\frac{\Delta E}{hc} = \frac{\nu}{c} = \tilde{\nu} \tag{1.10}$$

(i.e., the energy of the vibrational transition in wavenumbers is the same as the molecule's vibrational frequency).

In reality, molecules are not perfect harmonic oscillators. In other words, the variation in the potential energy of the system with internuclear separation is not perfectly symmetric, but rather tends to have a skewed appearance, similar to that of the *Morse potential*, as shown in Fig. 1.4. This type of potential energy dependence describes the behavior of an *anharmonic oscillator*, where the spacing between subsequent energy levels is reduced for higher-energy states [4].

Similar to the harmonic oscillator, Schrödinger's equation can be solved for the Morse potential, resulting in the following expression for energy states

$$E = h\nu\left[\left(\mathbf{v} + \frac{1}{2}\right) - x_e\left(\mathbf{v} + \frac{1}{2}\right)^2 + \cdots\right] \tag{1.11}$$

where x_e is called the *anharmonicity constant*. In general, higher-order terms in Eq. (1.11) are usually small and are routinely omitted. Energy-state separation between any two successive states can then be given by

$$\Delta E = h\nu[1 - 2x_e(\mathbf{v} + 1)] \tag{1.12}$$

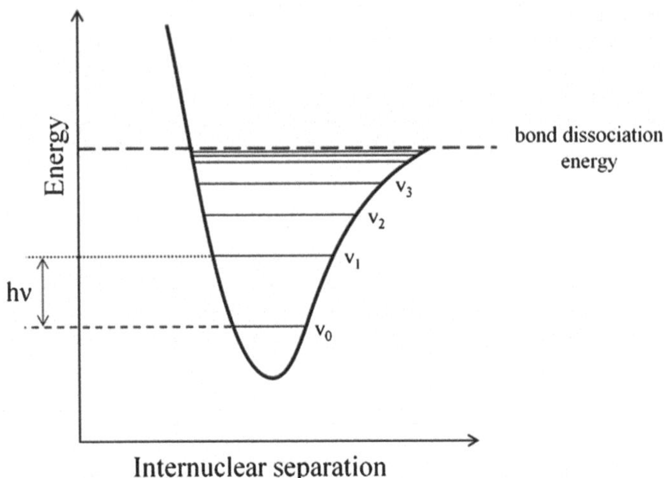

Fig. 1.4 Potential energy curve of an anharmonic oscillator (Morse potential). Energy-level separations decrease as **v** increases

where **v** is taken as the vibrational quantum number for the lower-energy state. In most cases, x_e is a positive number, so that the separation of successive states becomes progressively smaller as **v** increases.

The frequency recorded for a molecular vibration by IR spectroscopy corresponds to the energy difference between two vibrational states. The transition between $\mathbf{v} = 0$ and $\mathbf{v} = 1$ is known as the *vibrational fundamental*. For the IR absorption corresponding to the fundamental transition, we can see from Eq. (1.12) that the energy of the observed transition will be

$$\Delta E = hv[1 - 2x_e] \tag{1.13}$$

At the limit when $x_e \rightarrow 0$, one can easily see that harmonic oscillator behavior is restored (i.e., $\Delta E = hv$), and the observed vibrational frequency is the same as the molecular vibrational frequency.

Thus far, we have discussed harmonic and anharmonic representations of a diatomic molecule. Let us now consider the molecular motions of a polyatomic molecule with n atoms. The motions of each atom can be resolved into components along the three directions in the Cartesian coordinate system. Therefore, any molecule composed of n atoms has $3n$ *degrees of freedom*. These $3n$ degrees of freedom encompass vibrations, rotations, and translations. The vibrational motions of atoms can be expressed as fundamental vibrational modes of the entire molecule, known as *normal modes*. The number of vibrational modes will be $3n$ minus the number of non-vibrational modes. For linear molecules, which have three translational and two rotational motions, they possess $3n$-5 normal modes. In contrast, nonlinear molecules possess three rotational and three translational motions and hence have $3n$-6 vibrational modes. Apart from the general classification of stretching and bending modes, the bending modes can be further specified as scissoring, rocking, wagging, and twisting vibrations. The different kinds of IR-active vibrations are illustrated in Fig. 1.5 for the example of a CH_2 moiety.

1.3.2 Biomolecular Vibrations

As can be visualized from Fig. 1.5, along with the total number of molecular normal modes, it is easy to see that biomolecules can contain a large number of IR-active vibrational modes. Since most biomolecules contain a permanent dipole moment and have low symmetry, most normal modes are IR active. The stronger, measurable vibrational modes can lend key insights into molecular structure (i.e.,

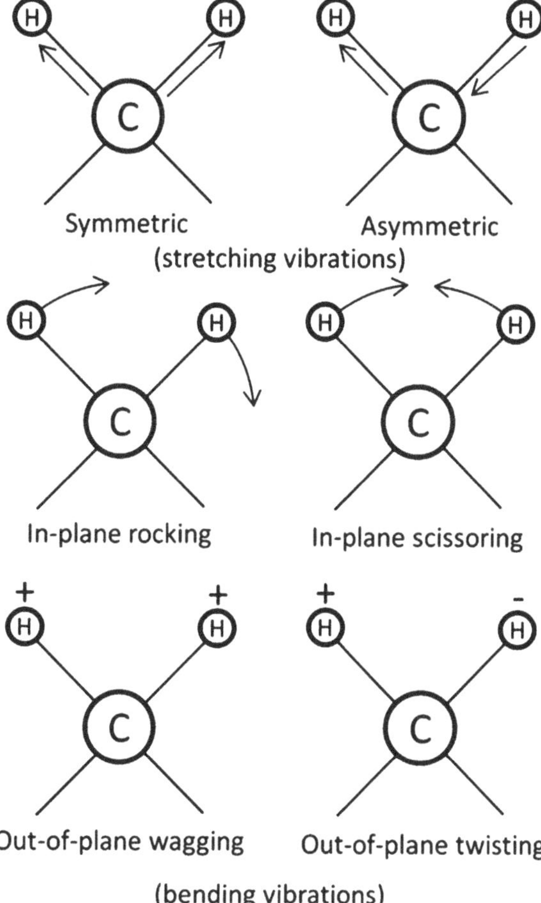

Fig. 1.5 Types of molecular vibrational motions shown for a CH_2 moiety

Table 1.2 Vibrational ranges for common biomolecular chemical moieties

Description	Frequency/cm^{-1}
P-OH stretch/P-OH wag	920–1,080
Amide NH bending	1,475–1,525
Amide CO stretch	1,675–1,725
Carboxylic acid CO stretch	1,725–1,775
Symmetric NH_2 stretch	3,350–3,400
Asymmetric NH_2 stretch	3,400–3,450
Amide NH stretch	3,300–3,500
Indole/Imidazole NH stretch	3,480–3,520
Carboxylic acid OH stretch	3,540–3,600
Alcohol OH stretch	3,600–3,675
Phosphate OH stretch	3,650–3,700

presence of chemical moieties) and typically reside within known frequency ranges. Some of the more biologically relevant vibrational modes are summarized in Table 1.2. With some of these vibrations lying in diagnostic regions of the electromagnetic spectrum, one can easily see why the IR spectrum is such a powerful tool in determining the chemical structures of analytes. Many of the remaining IR modes are either weak, and are hence challenging to measure, or located in congested regions of the IR spectrum, and are therefore difficult to unambiguously assign.

In the event that an experimental band assignment to a particular vibrational mode is ambiguous, chemical labeling strategies can be employed to deconvolute the IR spectrum. One such technique is *isotope labeling*, wherein an atom is selectively replaced with a heavier isotope of the same atom. For example, site-specific information can be obtained on an amide CO stretching vibration by replacing the ^{12}C with a ^{13}C.

Let us consider a diatomic CO molecule in the harmonic oscillator approximation. While this simplistic view is not an accurate representation of an amide moiety with full molecular considerations, the principle can be illustrated by looking at the vibrational frequency shift due to isotopic labeling of the carbon atom in the diatomic CO. Recall the frequency of a vibration given by Eq. (1.5), where the frequency is a function of the reduced mass of the system. When an atom is labeled with a heavier isotope, the reduced mass of the system is increased. Due to the inverse relationship between vibrational frequency and reduced mass, it is apparent that the vibrational frequency should be reduced upon an increase in the mass of the system (refer to Eq. 1.6). To illustrate this, the difference in frequency upon isotopic labeling is given by

$$\Delta \nu = \nu_{^{12}CO} - \nu_{^{13}CO} = \frac{\sqrt{k}}{2\pi}\left(\frac{1}{\sqrt{\mu_{^{12}CO}}} - \frac{1}{\sqrt{\mu_{^{13}CO}}}\right) \tag{1.14}$$

where the frequencies and reduced masses of interest are denoted by the respective carbon isotope labels (i.e., 12 or 13). The force constant of CO has been experimentally determined to be $k = 1{,}854$ N/m. Hence, a quick calculation yields a frequency redshift of $\Delta v = 48\ \text{cm}^{-1}$ upon ^{13}C labeling of a CO diatomic. Note that this frequency shift is the maximum possible frequency shift for a single oscillator. In larger molecules, a normal mode often involves oscillation of multiple subentities of the molecule. Due to the larger overall reduced mass, the effect of the isotopic substitution is markedly diminished, thus lessening the observed frequency shift.

1.4 Electronic Transitions

1.4.1 Background

The interaction between UV and vis light with matter generally entails an electronic transition (i.e., promotion of electrons from the electronic ground state [S_0] to a higher-energy state [e.g., S_1]) [5]. For a molecule, A, this process can be visualized by

$$A + hv \rightarrow A^* \tag{1.15}$$

where v is the frequency of radiation and A* represents an excited electronic state of A. Molecules containing π-electrons or non-bonding electrons can absorb energy from an UV or vis light source, which promotes electrons to higher-energy antibonding molecular orbitals. A schematic representation of the different types of transitions is given in Fig. 1.6. In general, there are three types of transitions; $\sigma \rightarrow \sigma^*$ transitions usually occur in the vacuum UV (i.e., <200 nm) and will not be considered here. The lower-energy $n \rightarrow \pi^*$ and $\pi \rightarrow \pi^*$, read "n to π star transition" and "π to π star transition," typically reside in the $200\ \text{nm} < \lambda < 750\ \text{nm}$ (UV–vis) region. The lowest energy transition is a promotion from the highest occupied molecular orbital (*HOMO*) to the lowest unoccupied molecular orbital (*LUMO*). For the example given in Fig. 1.6, the non-bonding orbital is the highest energy orbital that contains an electron and the π^* antibonding orbital is the lowest energy unoccupied orbital, which gives rise to the $n \rightarrow \pi^*$ HOMO–LUMO transition.

Another important aspect of electronic transitions is the fast timescale on which the absorption occurs (e.g., 10^{-15}–10^{-18} s). In general, electronic transitions occur several orders of magnitude faster than atomic vibrations ($\sim 10^{-13}$ s). According to the *Franck–Condon principle*, electronic transitions are considered instantaneous when compared to nuclear motions. In other words, absorptions, as shown on a potential energy curve, are essentially *vertical transitions* and the nuclear coordinates remain effectively unchanged during electronic excitations (see Fig. 1.7). Radiation of frequency v can be absorbed if the energy difference hv corresponds to the energy difference between a quantized energy level in the excited state and a quantized energy level in the ground state.

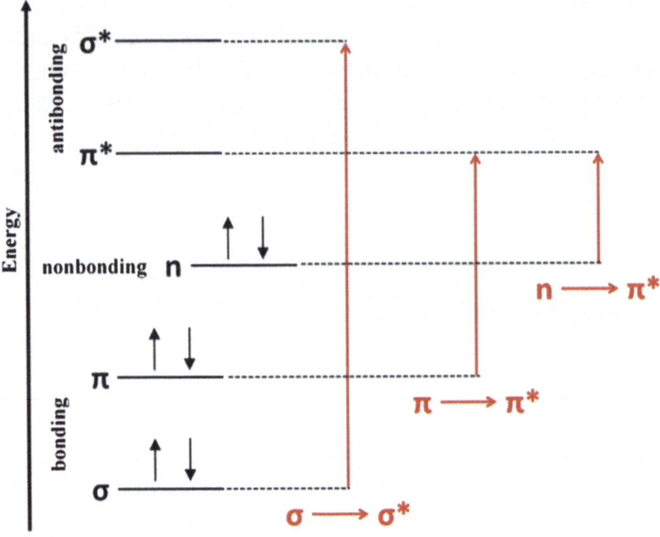

Fig. 1.6 Energy diagram depicting common electronic transitions

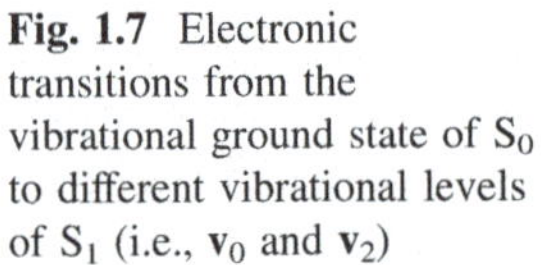
Fig. 1.7 Electronic transitions from the vibrational ground state of S_0 to different vibrational levels of S_1 (i.e., $\mathbf{v}_0$ and $\mathbf{v}_2$)

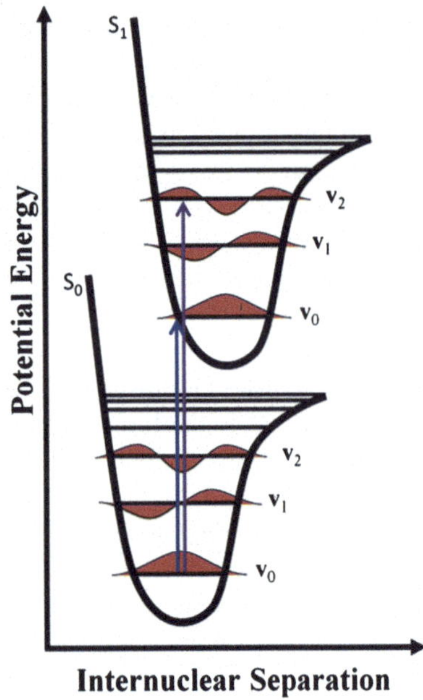

Since the electronic-state wavefunctions are orthogonal to one another (i.e., the overlap integral is equal to zero), the probability of an electronic transition is proportional to the square of the overlap integral between the ground-state

vibrational wavefunction and the excited-state vibrational wavefunction. The vibrational-state overlap integral is referred to as the *Franck–Condon factor* and governs the strength of electronic transitions based on the compatibility of the nuclear coordinates in the initial and final states of the transition. Hence, the probability of a transition is significantly reduced if the initial and final states correspond greatly to differing interatomic distances, as the nuclei would be required to undergo significant spatial changes accompanying the electronic transition. For the example illustrated in Fig. 1.7, one would expect the vibronic transition $\mathbf{v}_0 \rightarrow \mathbf{v}_2$ (purple arrow) to be stronger than the origin band $\mathbf{v}_0 \rightarrow \mathbf{v}_0$ (blue arrow), based on greater overlap, and a larger Franck–Condon factor, in the ground- and excited-state wavefunctions in the former case. For a more detailed discussion of the Franck–Condon principle, the reader is referred to relevant literature [3, 6].

1.4.2 Applications to Biomolecular Systems

For UV–vis spectroscopy to be applicable to biomolecular systems, a key requirement must be met. The system must contain a UV chromophore capable of absorbing photons in the UV–vis frequency range. For biological systems, this requirement is met in numerous cases. For instance, several amino acids act as UV chromophores (e.g., histidine, tyrosine, and tryptophan). Therefore, most systems (isolated amino acids, peptides, and proteins) that include these subunits will exhibit adequate UV absorption cross-sections. In addition, the peptide bond itself has the ability to absorb UV photons in the 190–230 nm range. On the other hand, saccharides do not exhibit absorption in the typical UV–vis range. Hence, such molecules must be subjected to chemical derivatization to be detected by UV spectroscopy. Since UV–vis spectra tend to contain superpositions of various vibronic transitions, and contributions from various conformers, absorption peaks tend to be broad and a limited amount of structural information can be obtained from these measurements. Some steps can be taken to sharpen the electronic transition bands and obtain structural (even conformer specific) information, a point that will be discussed further in Sect. 1.6.

1.5 Absorption Spectroscopy

To this point, we have discussed various types of transitions, both vibrational and electronic. Let us now turn our attention to the practical aspects of measuring the aforementioned transitions. First, we will consider the traditional absorption spectroscopy experiment. Absorption of photons by a species is governed by the Beer–Lambert law [5]

$$-\log(T) = A = \varepsilon lc \tag{1.16}$$

where T is the transmittance of light through the substance, A is the absorbance, ε is the molar absorptivity coefficient, l is the distance of interaction (i.e., the path length), and c is the molar concentration of the absorbing species. A simple measurement of T can be performed by measuring

$$T = \frac{I}{I_0} \tag{1.17}$$

where I and I_0 are the intensities of the incident and transmitted light, respectively. The UV–vis absorption spectroscopy experiment is illustrated in Fig. 1.8. Light from a light source is passed through a monochromator. The monochromatic light is split, passed through a blank absorption cell (i.e., to measure I_0), and passed through the sample (i.e., to measure I). A detector records the intensities of the respective signals, and a ratio of the two intensities is a direct measurement of T.

As can be seen in Fig. 1.8, a traditional absorption measurement is based on a simple setup, which, nonetheless, provides powerful measurements. From Eqs. 1.16 and 1.17, it is immediately apparent where the traditional absorption measurement becomes difficult. For this experiment to be successful, there must be a **measureable** difference between I and I_0. This becomes challenging when the values of ε, l, or c are small. Naturally, the value of l depends on the experimental setup; however, for practical reasons (e.g., instrumental cost, space requirements), it is best that the path length remains at a reasonable value. The value of c is sample dependent (e.g., sample availability, solubility), while the value of ε is inherent to the absorbing species.

In many physical chemistry experiments, the molar concentration of the absorbing species is extremely low. For instance, reactive species with short lifetimes may not exist in measurable quantities. In addition, absorption measurements are often ill-suited for the detection of gas-phase molecules, particularly ions in a mass spectrometer. To illustrate the limitations of the traditional absorption spectroscopy experiment, we consider the amino acid tryptophan in its ionized form. Tryptophan exhibits a moderate UV photoabsorption at 280 nm, with a molar absorptivity coefficient $\varepsilon = 5{,}690\ \text{M}^{-1}\text{cm}^{-1}$. If we reasonably assume an absorption path length $l = 1$ cm and a detection efficiency (I/I_0) of 0.1 %, then one can derive (from Eqs. 1.16 and 1.17) that the required concentration of tryptophan would need to be approximately 75 nM. This corresponds

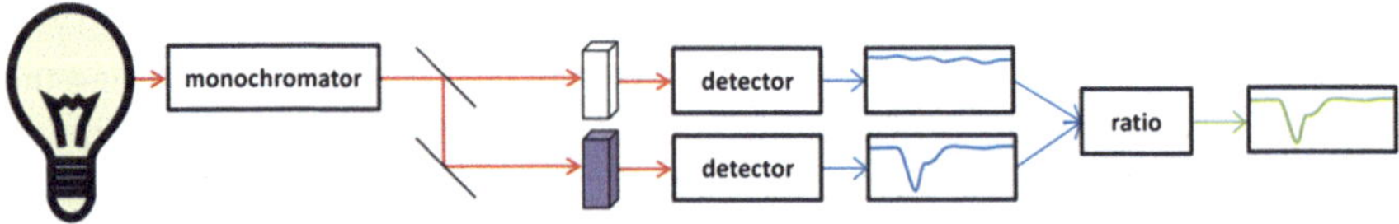

Fig. 1.8 Block diagram of a typical absorption spectroscopy measurement

to $\sim 4.6 \times 10^{13}$ tryptophan ions per mL of probed solution, which contrasts to a maximum of $\sim 10^7$ ions that can typically be trapped in a mass spectrometer. The vastly lower concentrations of ions in mass spectrometers clearly preclude spectroscopic interrogation via conventional absorption detection schemes. One must therefore turn to more sensitive spectroscopic methods—referred to as "consequence" or "action" spectroscopy—to enable these experiments.

1.6 Consequence Spectroscopy

1.6.1 Background

As illustrated in the previous section, the direct measurement of photon absorption imposes limitations that hinder experiments with low number densities of absorbing species. *Consequence* (or "action") *spectroscopy*, where it is not the absorption of photons that is detected, but rather the consequence of the absorption that is detected, overcomes these limitations. Consequence spectroscopic techniques are predicated in the fact that a measurable, physical property of the analyte molecules changes as a direct result of photon absorption. For the purposes of this book, as it relates to biomolecular ions, the mass, or *mass-to-charge* (m/z) *ratio*, is the physical property most often utilized for such measurements, as the m/z ratio is readily attained by a mass spectrometer (discussed further in Chap. 3). To illustrate, Fig. 1.9 schematically depicts the measurement of a consequence absorption spectrum via measurement of the mass spectrum of an analyte molecule. Analyte ions, of a given m/z ratio, are subjected to irradiation by a tunable light source (refer to Chap. 2) at discrete frequency steps, and changes in the m/z ratio are measured at each frequency step (Fig. 1.9 left). The m/z ratio of an analyte changes when a sufficient number of photons have been absorbed to cause photodissociation (i.e., generation of photofragments). As the frequency of the incident light is changed, the number of absorbed photons per ion, and thus the extent of photodissociation, changes according to the absorption cross-section at the given frequency. A consequence absorption spectrum can then be constructed by plotting the yield of photodissociation as a function of radiation frequency (Fig. 1.9 right). In this spectrum, molecular absorptions are manifested as an enhanced photodissociation yield, as opposed to no, or negligible, photodissociation at non-resonant frequencies.

Since this technique relies on the detection of ions with differing m/z values, rather than detection of the number of photons absorbed by an analyte, consequence spectroscopy techniques provide an indirect measurement of the absorption spectrum. Due to the indirect nature of the experiment, one must take caution not to overinterpret the results, as a number of assumptions are made. However, due to the sensitivity of modern-day mass spectrometers, the sensitivity of the measurement is telling. A mere, few hundred analyte molecules are sufficient, in stark

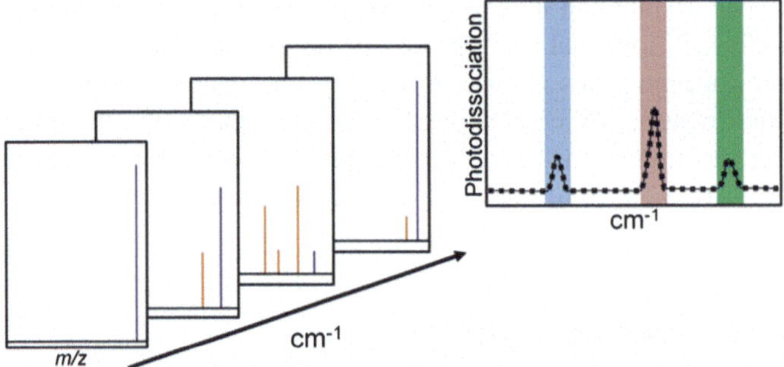

Fig. 1.9 Methodology of a photodissociation consequence spectroscopy measurement. Precursor ions (purple) photodissociate to fragments (orange) upon resonant photon absorption. Absorption bands are highlighted for illustrative purposes

contrast to the macroscopic number of analyte molecules required for a traditional absorption measurement.

Over the past few decades, a number of consequence spectroscopy techniques have been applied to analyze biomolecular ions, including (1) infrared multiple-photon dissociation (IRMPD) spectroscopy, (2) ultraviolet photodissociation (UVPD) spectroscopy, (3) infrared–ultraviolet (IR–UV) ion dip spectroscopy, (4) IR predissociation spectroscopy, and (5) photoelectron spectroscopy. The first four of these will be elaborated on with some detail. Note that IR–UV ion dip spectroscopy is an extension of UVPD, as it allows the measurement of IR spectra of UV-selected conformers. The conceptual approaches to IRMPD, IR predissociation, and IR–UV ion spectroscopy are summarized in Fig. 1.10, and each method will be discussed in turn below.

1.6.2 Infrared Multiple-Photon Dissociation Spectroscopy

IRMPD spectroscopy is easily implemented in commercially available trapping mass spectrometers. In terms of instrumentation requirements, IRMPD spectroscopy is the least technically demanding; however, this approach also results in the lowest resolution consequence IR spectra of the techniques that will be discussed. IRMPD requires only a trapping mass spectrometer and a tunable IR light source that can provide sufficient photon flux.

Generally, room-temperature ions are trapped for an extended period of time (i.e., seconds) and are irradiated with the output of a tunable IR photon source. The IRMPD process requires the absorption of multiple IR photons to surpass the dissociation threshold, visualized by a white dashed line in Fig. 1.10 left, of the analyte. When the IR photons are in resonance with the fundamental transition (i.e., $\mathbf{v}_0 \rightarrow \mathbf{v}_1$) of a vibrational mode, a photon is efficiently absorbed by the ions. However, subsequent transitions (e.g., $\mathbf{v}_1 \rightarrow \mathbf{v}_2$) are not resonant with the photon

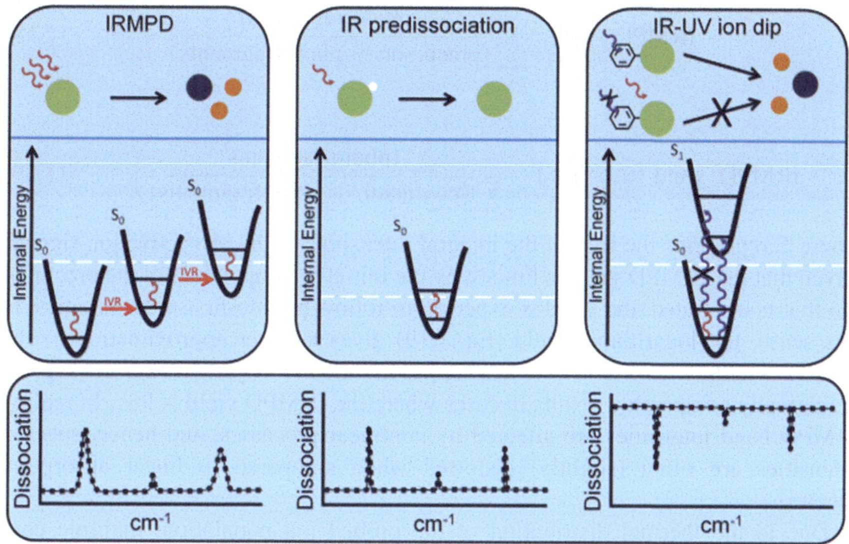

Fig. 1.10 Summary of infrared consequence spectroscopy techniques. Figure adapted from [7]

source, as the spacing between vibrational modes is reduced, due to anharmonicity. This "anharmonic bottleneck" prohibits dissociation via the so-called ladder-climbing process (i.e., $\mathbf{v}_0 \rightarrow \mathbf{v}_1 \rightarrow \mathbf{v}_2 \rightarrow \ldots \rightarrow$ dissociation). On the other hand, if other normal modes couple with the initially absorbing normal mode, energy can be dispersed throughout the molecule by *internal vibrational redistribution* (IVR). Through IVR, energy is quickly (i.e., ps timescale) dissipated throughout the bath of normal modes of the molecule, effectively depopulating the $\mathbf{v}_1$ state and returning the molecule to the vibrational ground state. The absorption of a subsequent photon at the fundamental frequency is then possible, again followed by dissipation of the energy throughout the molecule. This cycle is repeated many times (i.e., tens to hundreds), in the process raising the internal energy of the molecule to exceed the dissociation threshold of the molecule, and thus induces photodissociation into one, or multiple, fragmentation channels.

The IRMPD technique is based on a slow, sequential absorption of photons, causing a gradual increase in the internal energy (i.e., heating) of the ions. Therefore, in most cases, ions dissociate by the lowest energy dissociation threshold, such as the loss of a labile chemical moiety. In other words, higher-energy dissociation pathways are typically not accessible, due to the rapid randomization of energy that occurs via IVR. Moreover, while a particular chemical bond or moiety resonantly absorbs the photon energy, this group may not dissociate in the process, but rather other, more labile groups are involved in covalent-bond cleavage. The extent of photodissociation by IRMPD, or IRMPD yield, is quantified by

$$\text{IRMPD yield} = \frac{\sum(\text{photofragments})}{\sum(\text{precursor} + \text{photofragments})} \tag{1.18}$$

or

$$\text{IRMPD yield} = -\ln\left(1 - \frac{\sum(\text{photofragments})}{\sum(\text{precursor} + \text{photofragments})}\right) \tag{1.19}$$

where Σ represents the sum of the integral intensities of the respective ion signals. Given that the IRMPD yield is limited by the initial ion population of the precursor ion that is irradiated, the yield is expected to follow pseudo-first-order kinetics. In this sense, the logarithmic yield (Eq. 1.19) gives a better approximation of the relative yields, even if the linear approximation for yield (Eq. 1.18) gives numerically comparable results in cases where the IRMPD yield is low. In reality, IRMPD band intensities are affected by nonlinear processes, and hence, spectral intensities are often (slightly) distorted when compared to linear absorption spectra.

Due to the thermal distribution of the probed ion population, multiple conformers (e.g., rotamers) of the analyte ions are usually present, each with a slightly different absorption spectrum. Measured vibrational features in an IRMPD spectrum result from a superposition of multiple conformers, accounting for significant broadening (>15 cm^{-1}). An additional broadening effect arises from ion heating during multiple-photon absorption, as hotter ions have redshifted and broader absorption bands. In summary, the band-broadening effects in IRMPD are a result of both the conformational envelope (accessible at a particular temperature) prior to laser irradiation and the anharmonic effects (i.e., redshifting, broadening) during laser excitation. IRMPD spectroscopy is often well positioned to address questions on the chemical structures of ions, such as confirming the presence (or absence) of chemical moieties in analytes. However, the technique is less suitable to interrogate gas-phase conformations, for which other consequence spectroscopy techniques, which probe cold, conformationally restricted ions, are necessary.

1.6.3 Ultraviolet Photodissociation Spectroscopy

Due to the higher photon energy in the UV region of the electromagnetic spectrum, absorption of a single is generally sufficient to overcome the dissociation threshold. Photodissociation can take place by a number of routes, and the three most common pathways will be briefly discussed. In the highest energy mechanism, the UV photon absorption causes a *vibronic transition* (i.e., a simultaneous change in electronic and vibrational states) from the electronic ground state (S_0) to a highly excited, unbound vibrational level (i.e., above the dissociation threshold) in the electronic excited state, S_1. Promotion to a highly excited vibrational state, beyond the dissociation threshold of S_1, directly results in dissociation; this is shown in

Fig. 1.11 (left). The second mechanism differs only in that dissociation proceeds from the ground electronic state, following an *internal conversion* (i.e., electronic energy is converted into vibrational energy) from the electronically excited state. This can occur by

a. internal conversion directly to a highly excited vibrational state (i.e., above the dissociation threshold) of S_0
b. internal conversion to S_1 v_0, followed by relaxation to a highly excited vibrational state of S_0

and is shown schematically in Fig. 1.11 (center). The third mechanism proceeds via a potential energy curve crossing to a *dissociative state* (DS), which has no, or only a shallow potential energy minimum, hence causing dissociation (Fig. 1.11 right). This results in so-called predissociation or dissociation below normal dissociation thresholds. Electronically excited molecules may also be deexcited, without undergoing photodissociation, by the emission of a photon (i.e., *fluorescence*) or emission of an electron, depending on the energy of the incoming photon, and the charge state and polarity of the ion.

Naturally, UVPD experiments require a coupled tunable UV light source and a trapping mass spectrometer. A UVPD spectrum is measured by monitoring the photodissociation yield as a function of incident UV photon frequency. The relatively high-energy nature of the photons in the UV region permits single-photon (i.e., linear) dissociation. While structural information can be obtained from the electronic spectrum of a room-temperature ion, UVPD spectroscopy tends to result in broader spectral features, due to the multitude of available vibronic transitions

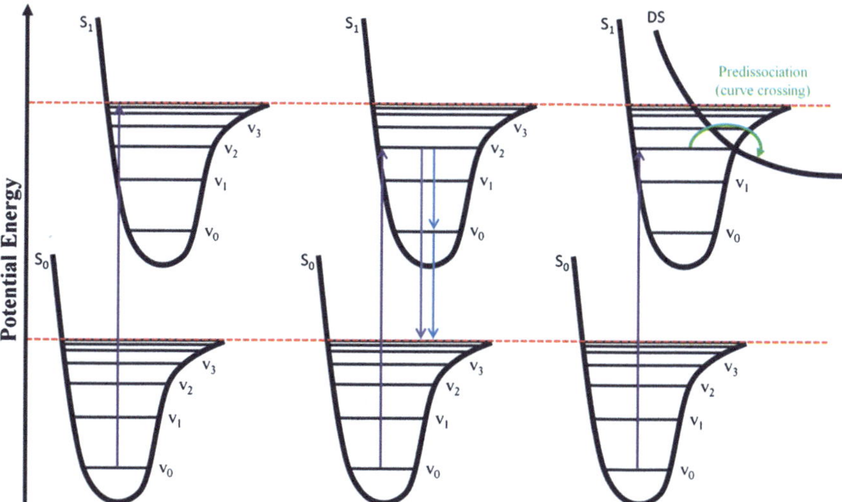

Fig. 1.11 Three potential dissociation mechanisms upon absorption of a UV photon. Dissociation thresholds are indicated via red, dashed lines

that are available. Similar to IRMPD spectroscopy, UVPD spectroscopy on room-temperature ions is more suitable for elucidating chemical structures, as opposed to characterizing conformations.

1.6.4 Infrared–Ultraviolet Ion Dip Spectroscopy

To obtain higher-resolution IR spectra than is possible by UVPD spectroscopy, a key detail must be addressed—the internal energy of the analyte molecules must be decreased to reduce conformational inhomogeneity. High-resolution UV spectra can be collected on cryogenically cooled ions, where sharp transitions correspond to origin and vibronic bands from a population of gas-phase conformers. In order to assign structures to specific UV transitions, it is useful to record the corresponding IR spectra. In IR–UV *ion dip spectroscopy*, conformer-specific IR absorption spectra can be collected by employing a two-laser photodissociation scheme.

Trapped ions are cooled to cryogenic temperatures prior to interrogation by IR–UV spectroscopy. In addition, IR–UV requires the use of a pulsed UV light source to photodissociate specific molecular conformers trapped within a mass spectrometer. Since the electronic (UV) transitions are intimately coupled with molecular structure, any two conformers will have slightly differing UV absorptions. The premise of the technique rests on the assumption that a particular UV transition is selective for a specific gas-phase conformation. As is implied, this technique is only applicable to analyte molecules that contain a chromophore capable of absorbing UV photons.

To obtain conformer-specific IR spectra, the UV light source is fixed to a unique UV transition, causing constant photodissociation of the selected conformer. Meanwhile, a tunable, pulsed IR light source is scanned to probe the analyte molecules. If the IR pulse precedes the UV pulse (e.g., by approximately 100 ns), and the IR frequency is in resonance with a $\mathbf{v}_0 \rightarrow \mathbf{v}_1$ vibrational transition of the conformer, an IR photon is absorbed, effectively depleting the ground vibrational state (S_0 $\mathbf{v}_0$). In a population of ions, a significant depletion of the ground vibrational state means fewer analyte molecules are in resonance with the UV photon frequency, therefore reducing the amount of measured UVPD (i.e., a "dip" in measured UV photofragmentation yield). In other words, as the frequency of the IR photons is scanned, each dip in the measured UV dissociation corresponds to a different vibrational transition in the molecular conformer being probed by the UV laser. This is visualized graphically in Fig. 1.10 (right). The IR spectrum can be recorded for multiple UV transitions, each potentially corresponding to a different molecular conformer.

IR–UV ion dip spectroscopy is the most sophisticated of the techniques mentioned here. In addition to the mass spectrometer and a cryogenic cooling apparatus, two light sources are required (UV and IR), both of which must be tunable and pulsed. The utility of these types of experiments can be witnessed in the high-resolution, conformer-specific absorption spectra that are generated (illustrated further in Chap. 4).

1.6.5 Infrared Predissociation Spectroscopy

Another spectroscopic technique that utilizes cold ions is IR *predissociation spectroscopy*, which uses a "messenger" or "tag" species (e.g., Ar, He, H_2) to report on photon absorption and requires that analyte molecules be cooled to cryogenic temperatures so that the messenger species can bind non-covalently prior to laser irradiation. A careful choice of tag species ensures that (1) the absorption bands of the analyte are not perturbed significantly by the presence of the tag and (2) the analyte molecules are sufficiently cold prior to binding, due to the low binding energy between the analyte and tag molecules. Cryogenically cooled analyte molecules have limited internal energy, reducing the number of possible conformations, and hence the spectral complexity. In addition, the low binding energy between the tag and analyte ensures that a single photon causes photodissociation, due to the dramatically reduced dissociation energy threshold (see Fig. 1.10 center). The linearity of this method also allows for the direct comparison of experimental results with *ab initio* electronic structure calculations, which are heavily relied upon to deduce structural information from experimental data.

As one may surmise, IR predissociation spectroscopy is more technically demanding than IRMPD spectroscopy, as in addition to a tunable IR light source and a trapping mass spectrometer, a means is required to cool analyte molecules to cryogenic temperatures. However, significantly higher resolution can be obtained, with feature bandwidths circa a few cm^{-1}. In addition, the predictable nature of the photodissociation channels (i.e., the loss of the tag molecule) allows the experiment to be multiplexed. In contrast to IRMPD and IR–UV ion dip spectroscopy, where numerous photofragments can be generated, predissociation generally results in the loss of a single tag molecule without subsequent dissociation. If overlaps in analyte *m/z* do not occur, in principle, the absorption spectra of several molecules can be measured simultaneously, dramatically increasing data collection efficiency (discussed further in Chap. 4).

References

1. Lide DR (2011) CRC handbook of chemistry and physics: a ready-reference book of chemical and physical data. CRC Press, Boca Raton
2. Giancoli D (2005) Physics: principles with applications. Upper Saddle River, Pearson/Prentice Hall
3. Atkins P, de Paula J (2006) Physical chemistry, 8th edn. W.H. Freeman, New York
4. Carter R (1998) Molecular symmetry and group theory. J. Wiley, New York

5. Harris D (2007) Quantitative chemical analysis. W.H. Freeman and Co, New York
6. Laidler K (1987) Chemical kinetics. Harper & Row, New York
7. Stedwell CN, Galindo JF, Roitberg AE, Polfer NC (2013) Structures of biomolecular ions in the gas phase probed by infrared light sources. Ann. Rev. Anal. Chem. 2013. 6:267–285.

Chapter 2
Light Sources

Nathan P. Roehr and Nicolas C. Polfer

2.1 Laser Theory

2.1.1 Light Sources Employed in Photodissociation Experiments

Photodissociating ions necessitate high-intensity light sources. Moreover, to record photodissociation "action" spectra on ions (see Chap. 1) requires the output wavelengths of these sources to be tunable. This chapter is not meant to give a comprehensive review of all light sources, but rather focus on the theory and operation of some of the benchtop light sources that have been successfully implemented in photodissociation experiments on ions. Figure 2.1 illustrates the overlap of these light sources with the vibrational and electronic regions of the electromagnetic spectrum. These include:

- Gas discharge lasers (e.g., CO_2 laser)
- Excimer lasers (e.g., ArF, KrF)
- Solid-state lasers (e.g., Nd:YAG, Ti:Sapphire)
- Nonlinear optics (e.g., OPO/A)
- Dye lasers
- Quantum cascade lasers QCL.

Other light sources, such as free-electron lasers (FELs), Raman shifters, fiber lasers, are not discussed in detail here. Note that some of those light sources are either not benchtop systems (e.g., FELs) or more suitable alternatives exist as light sources for photodissociation experiments of ions.

N. P. Roehr (✉) · N. C. Polfer
Department of Chemistry, University of Florida, Gainesville, FL 32611, USA
e-mail: nroehr@ufl.edu

N. C. Polfer
e-mail: polfer@chem.ufl.edu

N. C. Polfer and P. Dugourd (eds.), *Laser Photodissociation and Spectroscopy of Mass-separated Biomolecular Ions*, Lecture Notes in Chemistry 83,
DOI: 10.1007/978-3-319-01252-0_2,

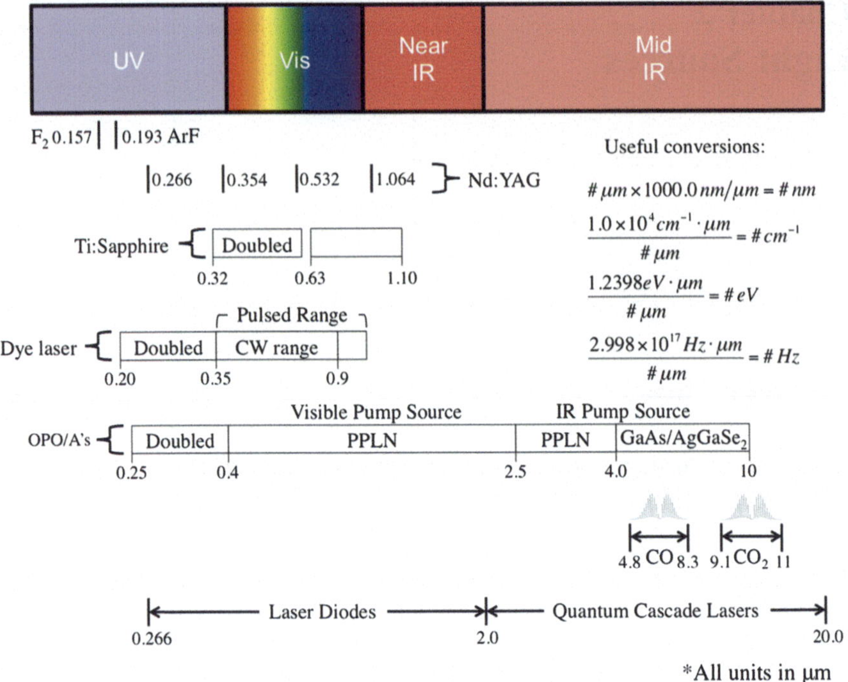

Fig. 2.1 Light source overview with all units given in μm [1–8]

In popular use, the term "laser" is often employed for intense light sources to carry out photodissociation spectroscopy experiments. It should be noted that while all lasers are light sources, not all light sources are lasers. However, because all light sources used for photodissociation experiments either are lasers or at least involve (pump) lasers, understanding lasing theory is useful for implementing these techniques.

2.1.2 Necessity of High-Intensity Light Sources

In 1969, Brauman and Smyth conducted and recorded the first successful photodissociation experiment. For a "tunable" light source, they used six optical filters and a 500 W tungsten bulb projector. Normally, high-intensity light sources are needed to carry out photodissociation experiments. To understand why high-intensity, laser-like, light sources are employed, consider photodissociation of the protonated tryptophan ion, $TrpH^+$. We assume the photodissociation threshold as the energy difference between the transition state (along the fragmentation pathway) and the ground state of a molecule. The $TrpH^+$ ion undergoes dissociation through a 8,160 cm^{-1} $molecules^{-1}$ (i.e., 23.3 kcal mol^{-1}) transition state [9]. A $TrpH^+$ ion will dissociate upon absorbing the following: a single photon from the UV, Vis, or near-IR >8,160 cm^{-1}; at least 3 photons from the hydrogen-stretching region ~3,000 cm^{-1}; at least 5 photons from the amide stretching region ~1,700 cm^{-1}; and >10 photons from the fingerprint region ~800 cm^{-1}.

Recalling from Chap. 1, the following equation describes the Beer–Lambert law:

$$-\log\left(\frac{I}{I_0}\right) = \varepsilon lc \tag{2.1}$$

In gas-phase spectroscopy, it is more convenient to use the following definition for the Beer–Lambert law:

$$-\ln\left(\frac{I}{I_0}\right) = \sigma lN \tag{2.2}$$

where σ is the absorption cross-section (cm^2 $molecule^{-1}$), l is the path length (cm), N is the number density (molecules cm^{-3}), and $\ln(I/I_0) = 2.303 \times \log(I/I_0)$. Rewriting Eq. (2.2) in a more convenient form, and rearranging the terms, yields:

$$\frac{I}{I_0} = e^{-\sigma lN} \tag{2.3}$$

$$1 - \frac{I}{I_0} = 1 - e^{-\sigma lN} \tag{2.4}$$

$$\frac{I_0 - I}{I_0} = 1 - e^{-\sigma lN} \tag{2.5}$$

The latter relation gives the ratio of photons absorbed. Strong absorption transitions, like those found in the UV or Vis, have σ values $\sim 10^{-16}$–10^{-17} cm^2 $molecule^{-1}$; for the weaker transitions, similar to those in the IR, σ values $\sim 10^{-18}$–10^{-20} cm^2 $molecule^{-1}$ are expected. Choosing nominal values for l and N (0.1 cm, 10^8 molecules cm^{-3}, respectively), and 1 s irradiation time, yields ratios of $\sim 10^{-9}$–10^{-10} in the UV/Vis, meaning that merely 1 in 10^9 or 1 in 10^{10} photons are absorbed. This extremely low ratio of absorbed photons illustrates the challenge in employing absorption spectroscopy in these measurements. Conversely, we can approach this problem from the point of view of the ions. In order to induce UV/Vis photodissociation in an appreciable fraction of the ions, one would need a photon source capable of outputting close to $\sim 10^{16}$–10^{17} photons cm^{-2} s^{-1}, to make detectable fragment ion quantities. These outputs are achievable even with incandescent light bulbs, indicating that single-photon UV/Vis photodissociation is readily implementable. For IR vibrations, the corresponding cross-sections are much lower (by ~2 orders of magnitude), and, at least in infrared multiple-photon dissociation (IRMPD), the absorption of multiple photons is required to induce photodissociation. In these implementations, the photon flux needs to be substantially higher, requiring higher-intensity light sources (experimentally $\sim$4.3 $\times$ 10^{18} photons cm^{-2} s^{-1} for $TrpH^+$ at 3,325 cm^{-1} [10]).

Note that the back-of-an-envelope calculation above also neglects other factors that can attenuate the experimentally determined photodissociation yield. Non-dissociative relaxation pathways, such as fluorescence, collisional cooling, and radiative emission, de-excite the ion, thus reducing the photodissociation yield. Finally, larger molecules have many degrees of freedom that are non-dissociative, and that hence slow down the photodissociation kinetics; this phenomenon is also known as the *kinetic shift* [11]. Slower photodissociation kinetics often imply a reduced photodissociation yield, as the competing, non-dissociative relaxation pathways above become more important. In order to speed up the photodissociation kinetics, ions need to be activated to higher internal energies, far above the dissociation threshold, requiring absorption of many more photons. These latter considerations play an important role in the requirement of intense light sources to induce efficient photodissociation.

2.1.3 Amplification by Stimulated Emission

The acronym *LASER* stands for light amplification through stimulated emission radiation, and the 1964 Nobel Prize in physics was given to C. H. Townes, N. G. Basou, and A. M. Prokhorov for their pioneering work in this field. To understand the operation of a laser, it is best to consider a simple two-level atom or molecule. Figure 2.2 shows such a system, involving lower and upper energy levels, as well as different processes for photon absorption and emission.

In *stimulated absorption*, the system is promoted into the upper level. From here, there are three means for the system to return to the lower level. *Spontaneous emission* involves de-excitation from the upper level to the ground level, accompanied by the release of a photon of light in a random orientation. In non-radiative relaxation, the system returns to the lower level without releasing a photon; collision with another atom or molecule is a common non-radiative relaxation pathway. In *stimulated emission*, another photon interacts with the upper level, thereby causing de-excitation and hence release of a photon with the same orientation and phase as the incident photon.

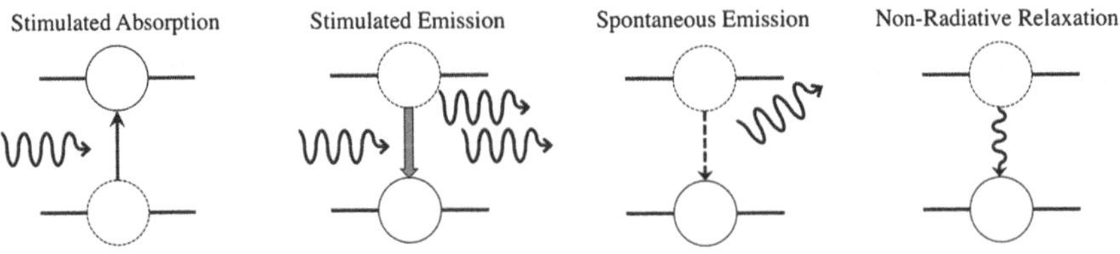

Fig. 2.2 Two-level representation of spontaneous absorption, stimulated emission, spontaneous emission, and non-radiative relaxation [12]

In a laser, a highly reflective optical *cavity* continually oscillates the light through the excited 2-level system, or gain medium, in order to amplify the process of stimulated emission. Figure 2.3 depicts the general schematic of a laser. A *pumping source* is employed to replenish the energy of the gain medium. The energy in the lasing cavity is stored in the form of a population inversion, where the excited state is considerably more populated than the ground state.

Only some percentage of the light is extracted from the laser cavity, as one of the mirrors is partially transparent; in Fig. 2.3, mirror B is partially transparent. In a *continuous wave* (cw) laser, the continuous loss of photons leads to a steady-state situation. Amplification has halted, leading to the more appropriate, yet simultaneously inappropriate, acronym of light oscillation by stimulated emission radiation (LOSER). In popular jargon, both pulsed and continuous wave light sources are designated as lasers.

There are two questions to consider: Why does lasing start initially and what causes stimulated emission to release a photon with the same orientation as the incident photon? Lasing is initiated by spontaneous emission of a photon that happens to be emitted in the direction of the laser cavity and is hence continually reflected through the gain medium. This photon then triggers stimulated emission of other photons, resulting in ever-increasing photon flux, in analogy to a chain reaction. In stimulated emission, the oscillating electric field of the incident photon interacts with the electrons in the excited species, so as to bring these electrons in phase with the oscillating field. This alignment rationalizes why photons with identical orientation and phase are emitted, effectively resulting in generation of coherent packets of photons.

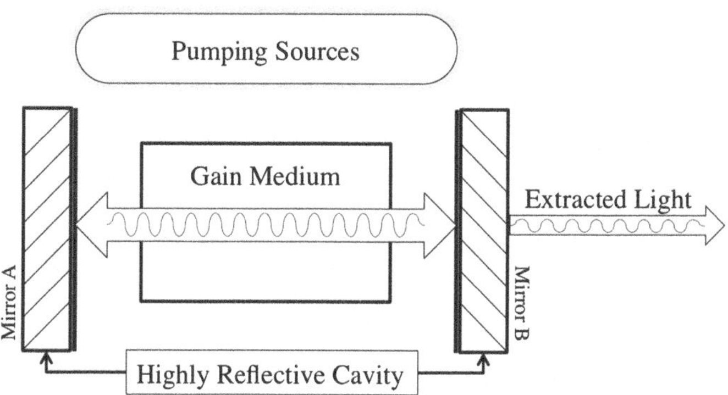

Fig. 2.3 Essential components of a laser [12]

2.1.4 Population Inversion

Figure 2.4 illustrates the importance of a population inversion for amplifying *photon flux* in a gain medium. N_1 is the population in the lower level, whereas N_2 represents the population in the upper level. The population N_2 must exceed the population of N_1 (i.e., $N_2 > N_1$), so that incoming photons trigger a cascade of emitted photons via stimulated emission. No gain can be achieved when the populations of N_1 and N_2 are equal, as stimulated absorption and emission cancel each other out. In fact, when $N_2 < N_1$, a net decrease in photon flux takes place, as stimulated absorption far outweighs stimulated emission.

Based on Fig. 2.4, the change in photon irradiance $\Delta I = I_f - I_i$ through the gain medium as a single pass can be approximated by

$$\Delta I = \sigma L I_i \Delta N \quad \text{where } \Delta N = \frac{N_2}{g_2} - \frac{N_1}{g_1} \tag{2.6}$$

where I_i is the initial photon irradiance (cm^{-2} s^{-1}), σ is the stimulated emission/absorption cross-section (cm^2), L is the length of the cavity (cm), N_1 and N_2 are the population number densities of the lasing transitions (cm^{-3}), and g_1 and g_2 are the degeneracies (i.e., number of states with identical energies). Let us assume equal degeneracies for both levels, as applied to all systems discussed in this chapter. For $N_1 > N_2$, the medium acts as an absorber. In contrast, when $N_2 > N_1$, the medium acts as an amplifier, as a result of a population inversion.

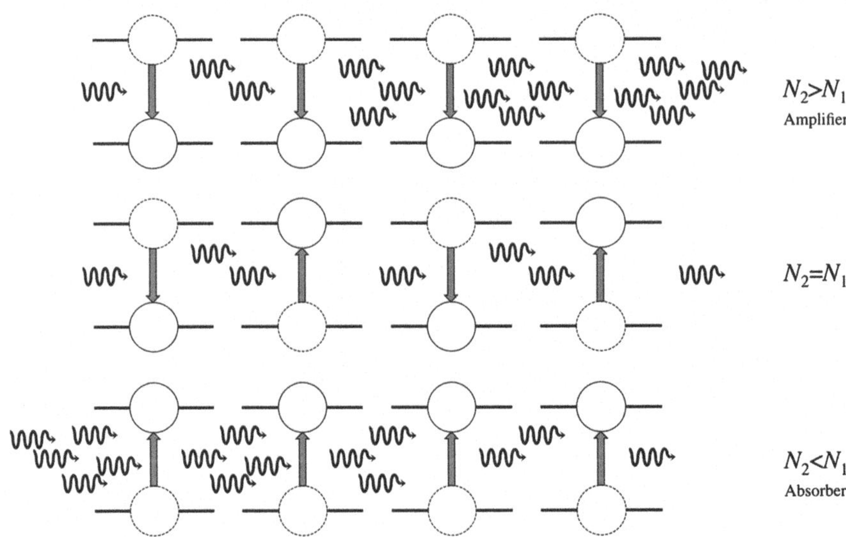

Fig. 2.4 Showing the cascading effect of stimulated emission and absorption for systems of different populations

2.1.5 3- and 4-Level Systems

A population inversion is typically not achievable in 2-level systems, precluding such systems for lasing purposes (the excimer laser being the exception). Conversely, in a 3- or 4-level system, a population inversion can be readily achieved, provided that the kinetic rates are suitable. Figure 2.5 depicts 3- and 4-level systems, where W_P is the pump rate, and γ_{ij} is the relaxation rate from level i to level j, respectively; the relative populations of N_i are also represented.

In the *3-level system*, a population inversion is achieved by a rapid decay from levels 2 to 3, but a slow decay from levels 3 to 1, i.e., $\gamma_{23} \gg \gamma_{31}$. This allows $N_3/N_1 > 1$. In a 4-level laser, pumping to level 2 (W_P) is followed by rapid conversion to level 3 (γ_{23}), then slower de-excitation to level 4 (γ_{34}) in the lasing transition, and again rapid conversion to the ground state (γ_{41}).

The *4-level system* is a much more efficient laser than the 3-level system. In the 3-level system, it takes considerable pumping to depopulate the lower level (i.e., level 1). Conversely, in a 4-level laser, level 4 may be initially unpopulated, thus allowing a facile buildup of a population inversion between levels 3 and 4. A comparison between a theoretical 3- and 4-level laser is represented graphically in Fig. 2.6. ΔN is plotted versus a normalized pump rate (W_P/γ_{31} and W_P/γ_{34} in a 3- and 4-level laser, respectively). It is clear from this graph that the lasing threshold condition, where gain in the cavity equals losses due to, e.g., absorption and scattering, is more easily met by the 4-level system. This is also true experimentally, as 4-level systems are largely more efficient than 3-level systems.

For either 3- or 4-level systems, significant spacing between levels is required to avoid heavy population, as a result of Boltzmann statistics. For instance, if

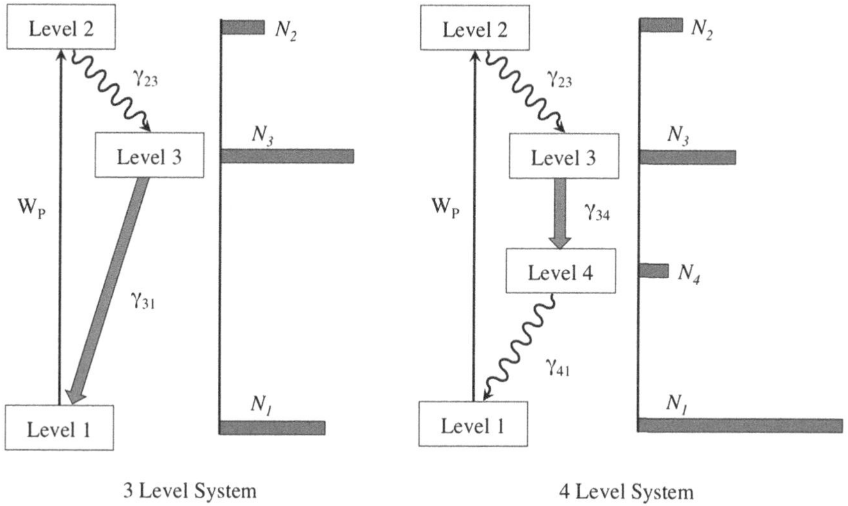

Fig. 2.5 3- and 4-level lasing systems with relative populations of levels. Adapted from [12]

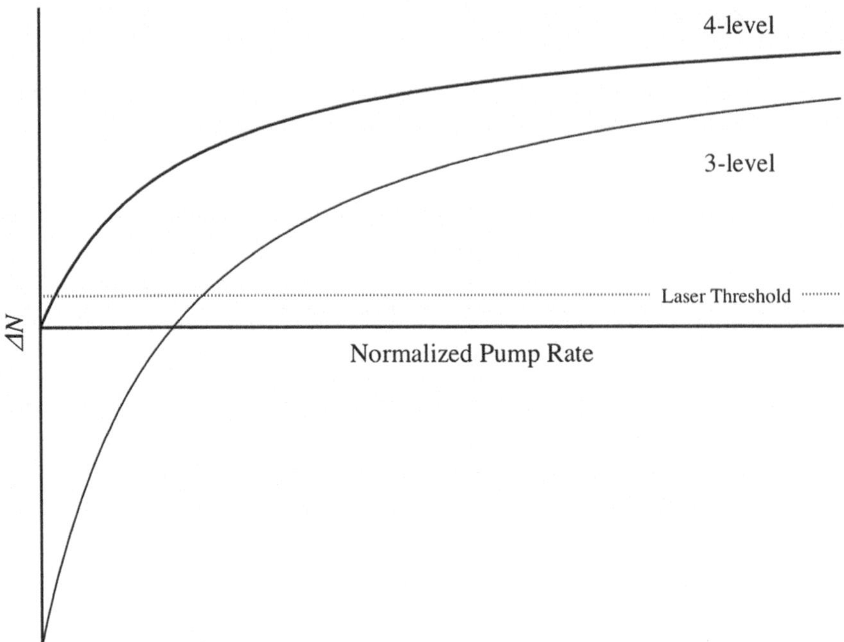

Fig. 2.6 Population inversion plotted versus normalized pump rate for a 3- and 4-level laser. Lasing threshold shown. Adapted from [12]

levels 1 and 4 in the 4-level system are separated by merely 200 cm^{-1}, the population ratio is at 300 K $N_4/N_1 \approx 0.4$. In other words, level 4 has an appreciable initial population, requiring a higher pumping power to achieve a population inversion between levels 3 and 4. If, on the other hand, the separation between levels 1 and 4 is 2,000 cm^{-1}, $N_4/N_1 \approx 7.3 \times 10^{-5}$ (at 300 K), and thus, level 4 essentially has no initial population.

2.1.6 Continuous Mode and Pulsed Mode (Q-Switching)

Lasers can be operated in two ways: cw and pulsed. In cw operation, the laser comes to a steady state, where *gain* = *output* + *losses*. This is shown in Fig. 2.3, where mirror B is partially transparent, continuously letting light escape from the cavity. In *pulsed mode*, light is extracted in bursts. This allows controlled timing of laser operation and dramatically increases peak power. The high burst of power from a pulsed laser is advantageous in promoting fast photodissociation, and in inducing nonlinear effects (discussed in Sect. 2.4), to allow wider tunability of output wavelengths.

The most common way to pulse a laser is by manipulating light oscillation in the cavity. The *quality* factor, or Q factor, expresses how well a resonator works and is defined as:

$$Q = \frac{\text{Energy stored}}{\text{Energy Lost Per Cycle}} 2\pi \tag{2.7}$$

A low Q factor implies that most/all of the energy in the cavity is lost, whereas for a high Q factor, the losses are minimized. A so-called *Q-switch* can toggle between high and low Q factors by either enabling or disabling light to resonate in the cavity at particular times. When the Q-switch is set to a low Q factor, large population inversions can be attained, as enhanced stimulated emission is disrupted. When the Q-switch is toggled to a high Q factor, stimulated emission in the direction of the cavity is promoted (by virtue of multi-pass oscillation). This process is started by spontaneous emission and is continuously amplified by stimulated emission, aided by oscillation in the cavity. Due to the very low initial flux, there is typically a certain time delay before the population inversion is depleted away, which is complemented by a sudden spike in the photon flux. The Q-switch then returns to a low Q factor, in order to build up another population inversion for the next cycle. Figure 2.7 depicts the Q value, population inversion, and output power all plotted as a function of time.

The earliest method of Q-switching used a mechanically spinning prism inside a resonator cavity. Modern methods use active Q-switches; electro-optical Pockel cells and acousto-optics; or passive Q-switches like saturable absorbers. An alternative Q-switch, cavity dumping, keeps the Q value high with 100 % reflective mirrors and then changes one mirror to 0 %, thereby dropping the Q value and "dumping" the energy out of the cavity.

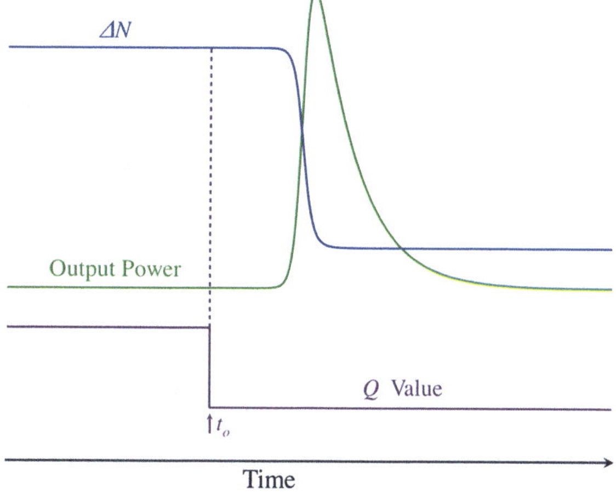

Fig. 2.7 Population inversion (ΔN), *output power*, and Q value as a function of time (arb. units) [13]

How a particular laser is pulsed can typically be found in its manual or by contacting the manufacturer. However, there are several important optical characteristics to consider when choosing a light source for a research project.

Repetition rate describes how often the laser can be pulsed and is typically given in pulses s^{-1}, i.e., Hz. The *pulse width* describes the full width at half max (FWHM) of the power versus time distribution of the output. The units are given in seconds. Integrating over the area of the peak gives *pulse energy*, in J, and the pulse energy/pulse width = *peak power*, in W. The *average power* of a pulsed laser is the amount of power in one cycle, i.e., pulse energy x repetition rate = average power. For cw lasers, there is only *output power* or *radiant power*, in *W*. The ideal optical characteristics naturally depend on the experiment.

2.1.7 Cavities and Tuning

Many lasers are capable of producing several discrete longitudinal modes or even a range of frequencies. To select a particular mode, a *diffraction grating*, *etalon*, *prism*, or *birefringent material* is incorporated into the cavity (e.g., see Fig. 2.8). These elements filter out unwanted frequencies and decrease the *spectral bandwidth*. The spectral bandwidth is an optical property describing the wavelength range output from a light source for a selected wavelength. No light source is truly monochromatic and will posses some spectral bandwidth.

A diffraction grating reflects light at a wavelength-dependent angle, a property due to the periodic grooves cut into the grating's surface. Figure 2.8 depicts a cavity in which a diffraction grating replaces one cavity mirror. At a particular tilt angle, the grating will allow cavity oscillation within a narrow bandwidth of light. In other words, when multiple lasing modes are accessible in the gain medium, the grating constrains amplification to one (or a few) mode(s) comprised in the selected bandwidth.

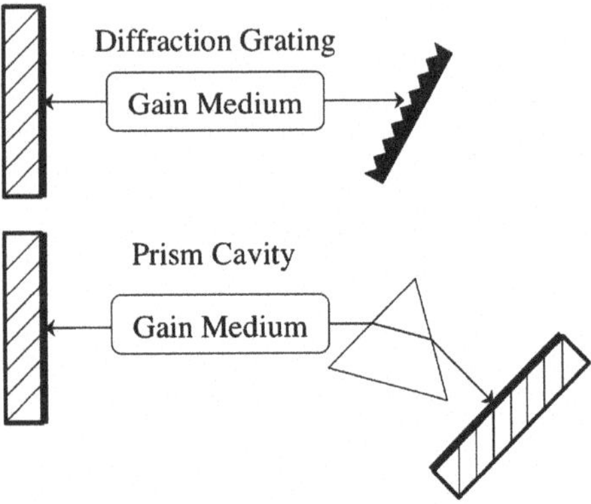

Fig. 2.8 Illustration of (*top*) laser cavity with diffraction grating, (*bottom*) laser cavity with prism

A prism acts similarly to a diffraction grating, except light is refracted at a wavelength-dependent angle. A mirror normal to the incident beam reflects the narrow bandwidth of light back through the prism to oscillate in the cavity. Frequency selection is achieved by changing the angle of incidence of the beam.

A breadth of cavity configurations for wavelength tunability exists, e.g., multiple prisms ring laser, etalon and diffraction gratings, multiple etalons and birefringent materials, each having advantages and disadvantages. As with the case of deciding on pulsed or cw mode light sources, these are important characteristics one must consider.

For an explanation of the Fabry–Perot interferometer (etalon) and Lyot filter (birefringent material) methods, the reader is directed elsewhere [2].

2.2 Gas Lasers

Gas lasers comprise any form of laser utilizing gas as a gain medium. Lasing transitions occur between discrete energy levels of electronic, vibrational, or rotational states of the gas species. Among gas lasers, the gas discharge CO_2 and CO cover a range of frequencies in the near-IR to mid-IR ranges, and the excimer laser has particular high-power UV output. The high power and routine operation make gas lasers very suitable candidates for photodissociation experiments of ions.

2.2.1 *The CO_2 Gas Discharge Laser*

The CO_2 laser belongs to the gas discharge family of lasers, which derive their name from the excitation method, namely a discharge between a cathode and anode in a gas-filled tube. The CO_2 laser has a high efficiency ($\sim$10–30 %) with output ranges between 9 and 11 μm.

Based on the $3n - 5$ relation for linear molecules, the triatomic (i.e., $n = 3$) CO_2 molecule in principle has 4 vibrational modes. However, due to symmetry considerations, the two bending modes are degenerate (namely ν_2, see below), thus only leaving three normal modes: Fig. 2.9 shows the symmetric stretching mode ν_1, the doubly degenerate bending mode ν_2 (note that only the mode involving oscillations in the plane of the paper is shown; the degenerate mode involves oscillations orthogonal to the plane of the paper), and the asymmetric stretching mode ν_3. The quantum vibrational state is written as $\nu_1\nu_2\nu_3$, where 000 indicates that all modes are in the ground state, and 001 denotes that the asymmetric stretching mode is in the first excited state, while all others are in their respective ground states.

The energy diagram for the CO_2 laser is shown in Fig. 2.10. The discharge excites an N_2 molecule to its first excited vibrational state. The N_2 molecule, as a homonuclear diatomic, lacks a transition dipole moment and can hence not relax

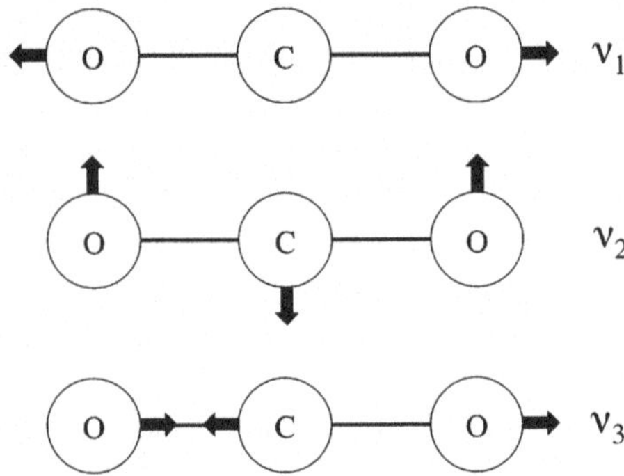

Fig. 2.9 The three normal vibrational modes of CO_2. Adapted from [13]

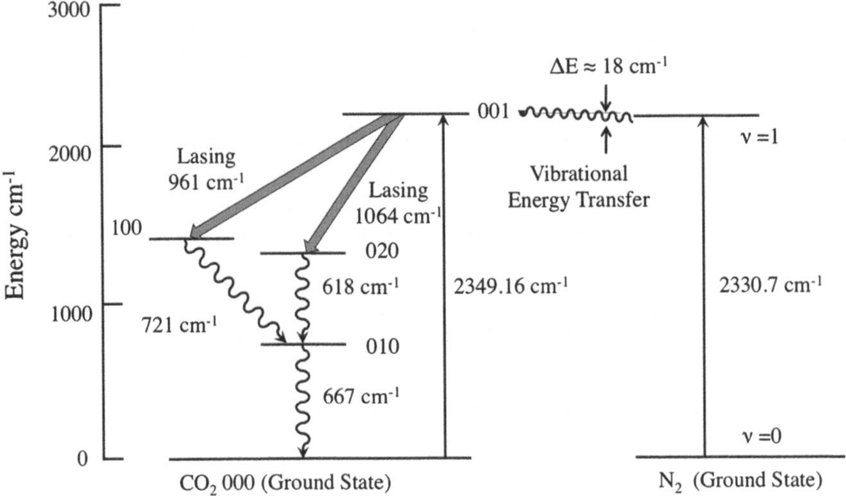

Fig. 2.10 Energy diagram of a CO_2 laser. Adapted from [13]

back to the ground state by spontaneous emission. Collisions between excited N_2 and ground-state CO_2 promote CO_2 to the first energy level of the asymmetric stretch, 001. Stimulated emission occurs from the 001 level to the symmetric stretching excited state 100 and the bending mode doubly excited state 020. These two intermediate energy levels relax via the 010 level to the ground state, enhanced by collisions with He atoms. This collisional relaxation occurs fast enough to deplete the 100 and 020 states, maintaining a population inversion for the lasing processes.

The optimal operation of a CO_2 laser requires a mixture of CO_2, N_2, and He (10–20 %, 10–20 %, remainder). The close spacing between the metastable, first excited N_2 vibrational state and the 001 state of CO_2 facilitates buildup of a population inversion. Collisional cooling by He is also essential to quickly depopulate the intermediate 100, 020, and 010 levels, in order to recycle the ground state. In addition, He maintains a steady discharge in the gain medium [6].

2.2.2 Rotational States and Rovibrational Transitions of CO_2

The lasing transitions are in fact rovibrational, which means that in addition to a change in the vibrational energy states, a change takes place in the rotational quantum number J. The rigid rotor is used to model these rotations with the quantized energies given by Eq. (2.8).

$$E(J) = \tilde{B}J(J+1) \quad \text{where } \tilde{B} = \frac{h}{8\pi^2 I} \cdot \frac{1}{c} \tag{2.8}$$

where $\tilde{B}$ is the rotational constant (cm^{-1}), I is the moment of inertia (kg cm^2), h is Planck's constant (kg cm^2 s^{-1}), and c is the speed of light (cm s^{-1}). These rotational levels are spaced more closely than vibrational levels (by ~1–2 orders of magnitude) and increase in spacing for higher J values (as apparent from Eq. 2.8). The selection rule prescribes that $\Delta J = \pm 1$. Figure 2.11 shows the rovibrational transitions of $\Delta J = +1$ (or the so-called P branch) and $\Delta J = -1$ (or the so-called R branch). The various rovibrational lines cover the 9 and 11 μm output range of a CO_2 laser. A particular transition is selected by virtue of a diffraction grating, as described in Fig. 2.8.

Using different isotopes of CO_2, e.g., ^{13}C or ^{18}O, can shift the laser output. The CO gas discharge laser is operated under similar conditions to a CO_2 laser, but due to the different rovibrational energy levels produces an output range from 4.8 to 8.3 μm [1].

2.2.3 Excimer Lasers and Other VUV Lasers

The high-power, high-efficiency excimer laser is the only 2-level system capable of lasing. Excimer lasers can only be operated in pulsed mode and have limited tunability (±1 nm). However, the high-intensity beams produced in the UV and vacuum UV (VUV) are excellent candidates for photodissociation and as pump laser sources. Table 2.1 lists output wavelengths of several excimer lasers. As a sidenote, the term excimer is derived from "excited dimer"; the dimers in question are made up of noble gas atoms, or noble gas and halide atoms.

The excimer laser is pumped with an electron beam or gas discharge. Pumping leads to the formation of metastable excimers, composed of dimers, as described in Table 2.1. These excimers have the unusual property of having bound excited states, but dissociative ground states. Figure 2.12 illustrates this concept for the example of the Xe_2 excimer laser. De-excitation from the excited Σ_u^+ to the ground

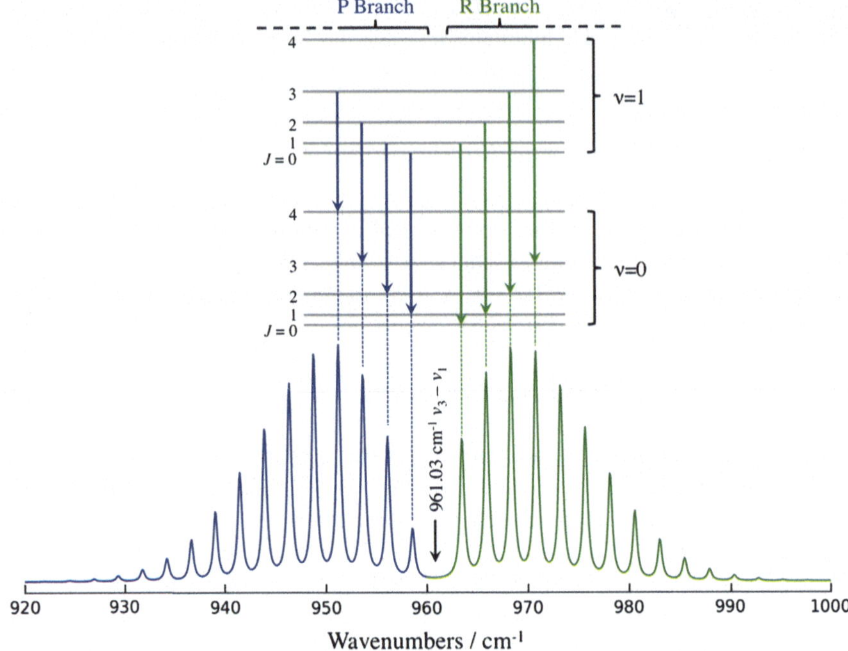

Fig. 2.11 *P* and *R* branch energy transition representation with corresponding simulated spectra

Table 2.1 List of several UV/VUV laser output wavelengths, excimer and F_2 [4, 14]

Laser	[a]F_2	Ar_2	Kr_2	Xe_2	ArF	KrF	XeBr	XeCl	XeF
Wavelength/nm	157	126	146	172	193	248	282	308	351

[a] not excimer

Σ_g^+ state leads to emission of a photon at 172 nm. The poor overlap between the excited Σ_u^+ and ground Σ_g^+ states leads to slow de-excitation for the lasing transition. The fast dissociation of the diatomic keeps the ground level Σ_g^+ depopulated with respect to the excited level Σ_u^+, thus maintaining a population inversion [15]. Although multiple steps precede the formation of the excimer species, only the two electronic states of the dimer participate directly in the lasing process, which is why these systems are known as 2-level lasers.

The formation of *exciplexes* (aka heteroexcimers), composed of diatomics of different atoms, such as ArF, KrF, or XeF, is accomplished by using a mix of halides and noble gases. The use of different excimer species allows a range of lasing wavelengths to be accessed [14], as summarized in Table 2.1.

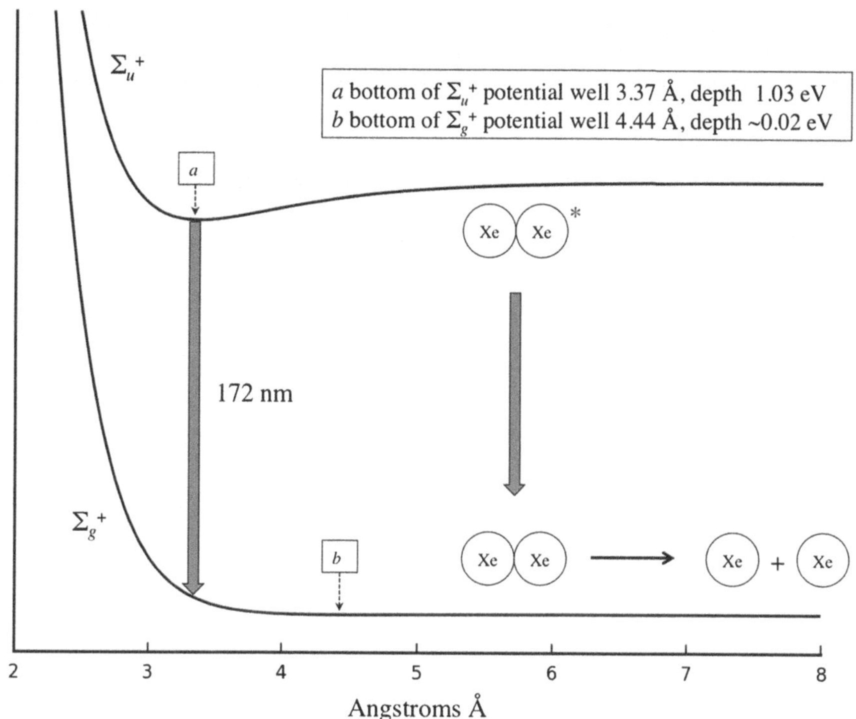

Fig. 2.12 Xe_2 energy diagram with lasing transition [14]

2.3 Solid-State Lasers

The original solid-state laser was a chromium-doped, pink *ruby laser* developed by T. Maiman in 1960. Solid-state lasers are prolifically used today, not only because of their wide tunability (some members), short wave pulses, and high energies, but also due to their robust construction and high gain factors. The gain medium of a solid-state laser is a glass or crystal doped with an activator ion. Transitions responsible for lasing action occur between inner electron clouds of the ion. Shielding by valence electrons allows the ion to almost behave like an isolated gas species, the only effect of the solid being Stark splitting of states, and in some cases, phonon interactions.

2.3.1 Nd:YAG

The neodymium-doped yttrium aluminum garnet laser, or Nd:YAG for short, is the most popular solid-state laser, finding its way in military, medical, and research applications. Being a 4-level laser, Nd:YAG has high efficiencies (some >50 %

[16]). The primary output is 1,064 nm with 35 other possible lasing transitions. Tunability using the sidebands of a transition is possible. However, their niche in photodissociation experiments is often that of a pump for other sources, e.g., for the widely tunable Ti:Sapphire laser and for optical parametric oscillators/amplifiers (OPOs/OPAs) (discussed below).

The $Y_3Al_5O_{12}$ garnet, doped with Nd^{3+} ions, is optically isotropic (i.e., uniform in all directions), thus minimizing loses. The garnet is also stable and hard, which is a useful property for manufacturing and long-term use. Finally, their good thermal conductivities allow for higher gain amplification.

Figure 2.13 depicts the energy transitions responsible for lasing action. The primary lasing transition at 1,064 nm occurs between split-levels in the $^4F_{3/2}$ state and $^4I_{11/2}$ state of Nd^{3+}. Lattice vibrations in the garnet facilitate depopulation of the $^4I_{11/2}$ state and relaxation from the pump bands to $^4F_{3/2}$ [2]. Broadening of the pump bands also occurs from lattice vibrations, so-called phonons, allowing a variety of pump sources. Early designs used flash lamps, but experiments requiring high spectral purity, like nonlinear optics, suffered from the lamps' instability [17]. The modern Nd:YAG laser is monochromatically pumped with a diode laser (see Sect. 2.6), resulting in higher spectral purity.

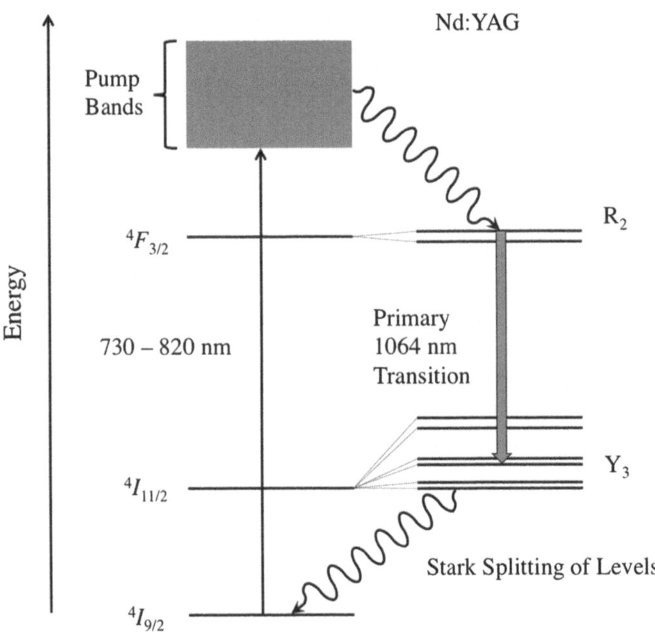

Fig. 2.13 Energy diagram of Nd: YAG laser, showing Stark splitting of levels. Adapted from [12]

2.3.2 *Widely Tunable Solid-State Lasers*

The tunability of solid-state lasers arises from the coupling of lasing transitions of the dopant material with lattice vibrations of the crystalline material. Figure 2.14 lists some of the more tunable solid-state lasers.

The prominent titanium-doped sapphire (Ti:Sapphire) laser is distinguished for tuning across the Vis to near-IR range. Initially stated to output 700–1,000 nm, some modern designs boast 630–1,100 nm output. The energy diagram is depicted in Fig. 2.15. The Ti:Sapphire is based on a 4-level scheme. Initial pumping promotes the Ti^{3+} ion from the ground vibrational level of $^2T_2$ to excited vibrational levels of the excited electronic state ^2E. Vibrational relaxation brings ions into the ground vibronic level of ^2E. From here, lasing action can occur to any of the $^2T_2$ vibronic levels—allowing wide tunability. Vibrational relaxation returns Ti^{3+} to the ground vibronic level. Pumping is usually accomplished with a doubled Nd:YAG laser or variant thereof, or alternatively an Ar-ion laser.

2.4 Nonlinear Optics Light Sources

When a light wave propagates through a medium, the wave's electromagnetic field interacts with the atoms in the medium. This interaction displaces the atoms' valence electrons, thereby creating dipoles. The dipoles affect the electromagnetic fields, giving rise to polarization effects that scale linearly with the electric fields. However, at higher field strengths nonlinear, higher-order polarization effects are observed. Equation (2.9) relates the magnitude of induced polarization to field strength.

$$\mathbf{P} = \varepsilon_0 \chi^{(1)} E + \varepsilon_0 \chi^{(2)} E^2 + \varepsilon_0 \chi^{(3)} E^3 + \cdots \tag{2.9}$$

where $\mathbf{P}$ is the induced polarization of light (C m^2), ε_0 is the free-space permittivity (F m^{-1}), E is the electric field (N C^{-1}), and $\chi^{(n)}$ denote the nth order susceptibilities (unitless). $\chi^{(1)}$ relates to linear effects, while the higher-order terms relate to

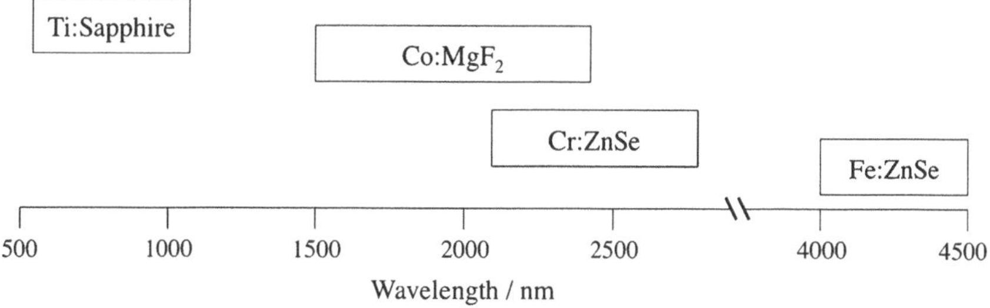

Fig. 2.14 Spectral range of some tunable solid-state lasers [18–21]

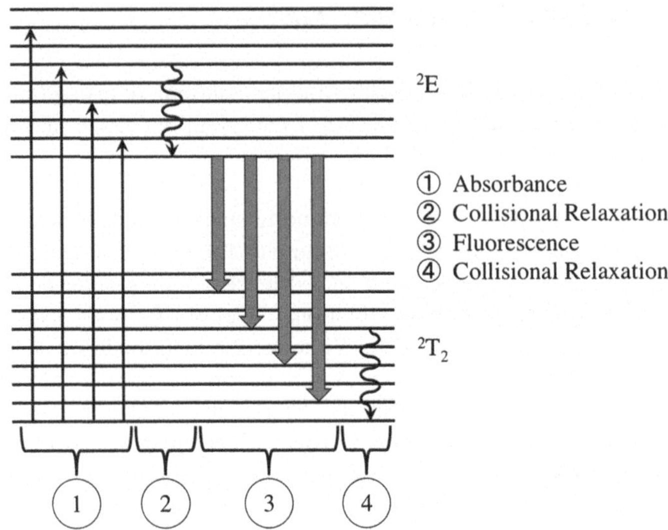

Fig. 2.15 Ti:Sapphire energy diagram. Adapted from [2]

nonlinear effects. The polarization arising from the second-order susceptibility, $\chi^{(2)}$, scales by the square of the electric field (i.e., E^2). $\chi^{(2)}$ effects can be employed for sum (i.e., $\nu_1 + \nu_2 = \nu_3$) and difference (i.e., $\nu_1 - \nu_2 = \nu_3$) frequency generation, and optical parametric generation/amplification. Suitable optical materials to access these effects must be non-centrosymmetric (i.e., they must lack an inversion center), since inversion symmetry operations cancel out even-order susceptibilities. Other properties necessary for nonlinear crystals include high transmissivity and a high optical damage threshold.

The most common example of second-order susceptibility is frequency doubling of the Nd:YAG output from 1,064 nm to 532 nm. Frequency doubling is a special case of sum frequency generation, where $\nu_1 = \nu_2$, and hence $\nu_1 + \nu_2 = 2\nu_1$. Using a second crystal, summing the second harmonic (i.e., 532 nm) and principal frequency (i.e., 1,064 nm) produces the third harmonic at 355 nm. Alternatively, doubling the second harmonic produces the fourth harmonic at 266 nm. In this way, nonlinear crystals can be used to produce output in the UV and Vis range with IR pump lasers.

2.4.1 Optical Parametric Oscillators/Amplifiers

For wider tunability in output, the optical parametric oscillators/amplifiers (OPOs/As) can be used. The basic setup of optical parametric generation (OPG) and OPA is depicted in Fig. 2.16. In nonlinear OPG, a pump photon of frequency ν_P enters the nonlinear medium and is annihilated as it excites a virtual state in the crystal.

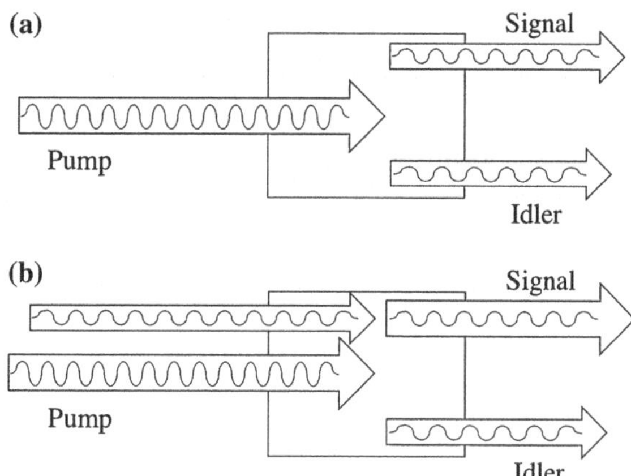

Fig. 2.16 Illustration of (**a**) optical parametric generation (*OPG*) and (**b**) optical parametric amplification (*OPA*) [2]

Relaxation from the virtual state releases two photons at different frequencies: the signal ν_S and idler ν_I frequencies. The three frequencies are related by the conservation of energy, where $\nu_P = \nu_S + \nu_I$. During optical parametric amplification, a pump and signal beam enter the medium. The pump is extinguished, in the process amplifying the signal and producing an idler photon. In generation and amplification, the pump, signal, and idler waves need to remain in phase with each other throughout the crystal, or lose intensity. This is called *phase matching* and is achieved when $n_P\nu_P = n_S\nu_S + n_I\nu_I$.

2.4.2 Nonlinear Crystals

Table 2.2 lists properties of crystals commonly used for nonlinear optic applications. *Potassium dihydrogen phosphate* (KDP) has a lower nonlinear coefficient than many of the other crystals, but a much higher damage threshold and can be grown to large sizes (a higher nonlinear coefficient means a higher nonlinear effect). *Beta-barium borate* (BBO) is effective at frequency doubling, but both KDP and BBO are hygroscopic and require hermetic sealing. *Potassium titanyl phosphate* (KTP) has a high nonlinear coefficient and is not hygroscopic, but has low damage threshold and can only be grown in small sizes. For the mid-IR region, the $AgGaSe_2$ nonlinear properties are frequently utilized. They have a high nonlinear coefficient and broad transparency, the only drawback being a lower damage threshold [22].

A method known as *quasi-phase matching* (QPM) allows the use of a higher nonlinear coefficient. This is accomplished by using periodically poled material, in other words, a non-centrosymmetric crystal with alternating symmetry inversions. *Lithium niobate* ($LiNbO_3$) when phase matched has an effective nonlinear coefficient of 4.7 pm/V, while periodically poled lithium niobate (PPLN) using QPM

Table 2.2 Physical properties of several nonlinear optic crystals [2, 22, 23]

	KTP	KDP	BBO	$AgGaSe_2$	PPLN	OP-GaAs
Transparency (μm)	0.35–4.0	0.22–1.6	1.9–3.3	1.0–18	0.35–5.2	0.7–17
Nonlinear coefficient d_{eff} (pm/V)	3.18[a]	0.37[a]	1.94[a]	32.4[a]	17[b]	86[b]
Damage threshold (GW/cm^2)	1	8	1.5	0.02	0.2	2
Nominal size (cm^3)	1–2	>30	10	15	0.002	10
Hygroscopic	No	Yes	Yes	No	No	No

[a] Phase matching
[b] Quasi-phase matching

has an effective nonlinear coefficient of 17 pm/V. PPLN and the recently discovered *orientation patterned GaAs* find their niche as OPO/A crystals [3].

For PPLN, the principal at 1,064 nm or second harmonic at 532 nm from a Nd:YAG is used to pump an OPO/A. Turnkey sources are available from ~250–2,500 nm for UV to near-IR applications (i.e., the ~250–500 nm range being accessible via frequency doubled output) and from 2,500–4,000 nm (i.e., 4,000–2,500 cm^{-1}) in the hydrogen-stretching region. GaAs has an extensive range of transparency down to 17,000 nm (588 cm^{-1}) and is tunable from 2.0 to 17 μm, provided that a pump laser <1.7 μm is available for effective conversion. More complicated OPO/A setups, using the idler from a PPLN OPO in a down-conversion process with other crystals are becoming more routine, even if the lower optical thresholds of the down-conversion crystal limit the output powers (i.e., ~μJ) [24, 25]. Note that these low powers are still ample to implement IR photodissociation spectroscopy of cryogenically cooled ions (as discussed in Chap. 1).

2.5 Dye Lasers

In 1966 and 1967, the dye laser came to prominence, filling the role of an easily tunable laser with access to a wide frequency range, producing output from the UV to the NIR. Further developments led to dye lasers with short pulse lengths (i.e., ns) and narrow spectral widths. It was the first light source to be employed in photodissociation experiments of biomolecules and is still used to this day for that purpose. Nonetheless, dye lasers suffer from a number of drawbacks vis-à-vis their solid-state competitors and nonlinear optics and are hence being displaced by OPO/A systems in some applications.

2.5.1 Dye Laser Theory

Figure 2.17 depicts characteristic eigenstates and transitions of dye lasers. (1) Stimulated absorption excites a molecule to the S_1 or S_2 states. Pumping is usually carried out with a flash lamp or another laser. (2) A fast (~10^{-11} s)

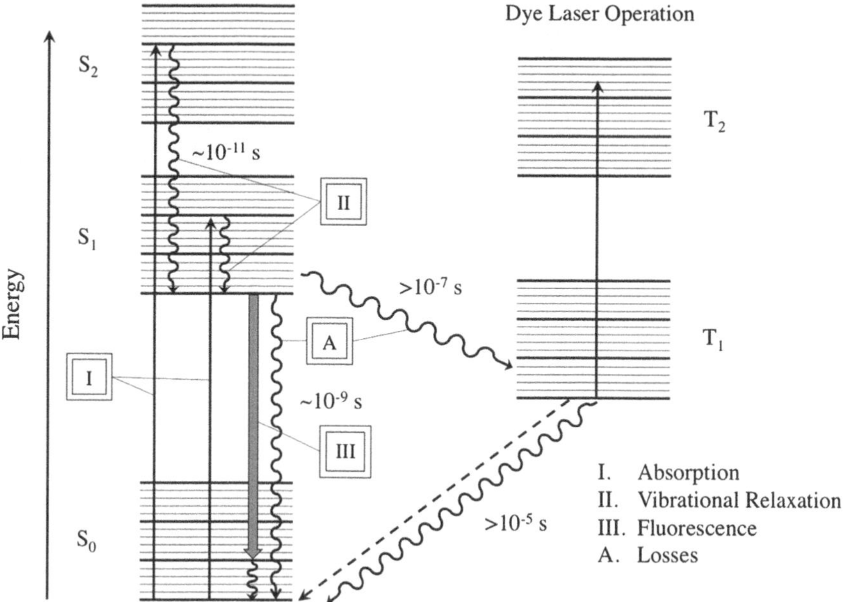

Fig. 2.17 Energy diagram for a typical dye laser. Adapted from [5]

relaxation brings the molecule into the ground vibrational state of S_1. (3) Stimulated emission to multiple vibrational states in S_0 can be mediated, making the dye laser widely tunable. Similarly to a CO_2 laser, specific transitions are selected by introducing a grating, grating-etalon, or prism(s) into the reflective cavity.

The excited vibrational states of S_0 are quickly ($\sim 10^{-11}$ s) depleted to the ground vibrational state, thus maintaining a population inversion. However, conversion from S_1 singlet to T_1 triplet ($>10^{-7}$ s) diverts a part of the excited state population away from lasing processes. Note that molecules in T_1 naturally return to the ground electronic state via the relatively slow process of phosphorescence ($\sim 10^{-6}$ s), but other processes, such as reactions, can lead to the destruction of the dye. Quenchers are added to accelerate the deactivation to S_0 and hence minimize the population in T_1 (or even T_2).

2.5.2 Dyes Used

Laser dyes exist for wavelengths between 340 and 1,180 nm, but only a few can be operated in cw mode. Though dyes supporting a tunability range of ~ 100–$200\ \text{cm}^{-1}$ exist, many dyes may be contained in a single cavity to cover a spectral region of interest, commonly the Vis. Table 2.3 shows the accessible ranges of dyes used for the visible region. Coumarin and rhodamine derivatives are

Table 2.3 List of dyes used [5, 26]

Dye	Lasing wavelength/nm
PBBO	386–420
Coumarin 120	428–453
Coumarin 102	460–495
Coumarin 153	515–575
Rhodamine 101	621–670
Pyridine 1	664–740
Styryl 8	738–794
Styryl 9	785–845

the most common, but others are also used, e.g., styryl, pyridine. Because lasing causes heating of the dye, and thus a loss in lasing efficiency, a circulation cooling system is employed.

2.6 Semiconductor Lasers

In terms of sheer numbers, semiconductor lasers are the most prolifically used, laser pointers in oral presentations being a notable example. Monochromatic semiconductor diode lasers are available from the UV to the near-IR with average output powers of up to Watts. Diode lasers make excellent pump sources for other lasers, particularly where excessive jitter from flash lamp pumping would overly complicate a setup or render it unfeasible. Recent advances in the widely tunable mid-IR quantum cascade laser (QCL) are about to bring the idea of benchtop turnkey mid-IR lasers to fruition.

2.6.1 Semiconductor Basics

As opposed to electrons in atoms or molecules, electrons in solid materials have bands of allowed energy that they can occupy, not discrete atomic or molecular orbital levels per se. In semiconductors, the lower-energy valence electrons denote the bound electrons, while the higher-energy conduction electrons can move freely around the lattice. Both levels are separated by a band gap, also known as inter-band gap or forbidden band. The properties of valence, conduction, and forbidden band are dependent on the semiconductor material that is used.

A process known as *doping*, where impurities are added to the semiconductor material, produces allowed energy levels within the forbidden band. Figure 2.18 schematically illustrates where the doped levels are inserted in the energy gap between the valence and conduction bands. In *n*-type semiconductors, electron donors are added, which results in filled energy levels *above* the valence band.

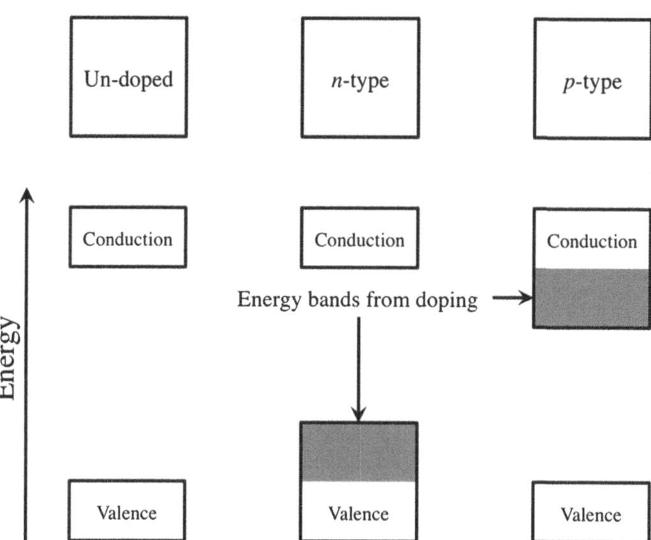

Fig. 2.18 Illustration of conductance and valence bands of undoped, *n*-type, and *p*-type material

Conversely, in *p*-type semiconductors, electron acceptors (i.e., positive holes) are added, which translates into available energy levels *below* the conduction band. The net effect from doping is that the interband gap is reduced. When *n*-type and *p*-type semiconductors are in contact at a junction, electrons can flow from the *n* to *p*-type, but are restricted from flowing from *p*- to *n*-type. The combined *np* device is called a semiconductor diode. As a sidenote, by convention, the direction of electrical current is in fact defined as the direction that the electron holes (i.e., lack of electrons) travel, which is in the reverse direction that the electrons migrate (i.e., electrical current is transmitted from *p* to *n* in a diode).

The interband gap is typically too large to be overcome by thermal excitation. For an electron to be promoted from the valence to the conduction band (or, in other words, a positive hole to be promoted from the conduction to the valence band), a photon equal in energy to the band gap must be absorbed. The latter process denotes interband absorption. In the reverse process, in interband emission, the relaxation of an electron from the conduction to the valence band is accompanied by the release a photon equal in energy to interband gap; the latter process is also referred to as electron-hole recombination.

2.6.2 Laser Diode Operation

Purposed in 1957 and realized in 1962, the diode laser utilizes interband emission for lasing action. Laser diodes are routinely manufactured as a single monochromatic source in the ~345–2,000 nm (to ~266 nm by frequency doubling).

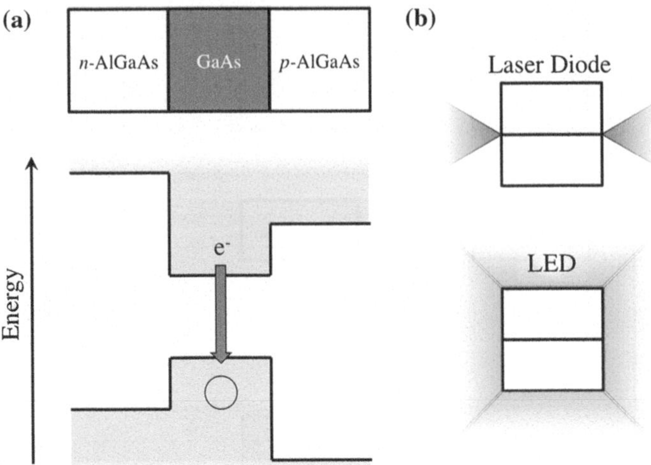

Fig. 2.19 **a** Energy diagram for GaAs diode laser, **b** output profile for laser diode and LED [13]

Figure 2.19a illustrates the basic components and energy diagram for a GaAs diode laser. Bracketing a thin sheet of GaAs between an *np*-type diode composed of AlGaAs creates a 4-level type system with GaAs's conduction and valence bands as the lasing levels 3 and 2, respectively. Electrons quickly flow into the GaAs conduction band from the *n*-AlGaAs conduction band and quickly flow out of the GaAs valence band into *p*-AlGaAs valence band. An electric current typically pumps the laser diode, although optical pumping is possible.

Light emission in laser diodes and light-emitting diodes (LEDs) rests on the same fundamental principle, with the notable difference that LEDs emit light spontaneously in all directions, while laser diodes are set up to oscillate light through the lasing medium, giving the output directionality (as shown in Fig. 2.19b). Since the output photon frequency is based on the interband gap, altering the interband gap, e.g., by doping, temperature control, or employing different materials, allows tuning of the desired frequency output.

Due to the broad distribution of electrons in the lasing media layer's conduction and valence bands, the laser diodes can have a broad bandwidth ($\sim$20–30 nm) output and, in addition, have large beam divergences. These undesirable features can be constrained using appropriate optical elements, such as filters and lenses.

2.6.3 Quantum Cascade Laser Operation

Bell Laboratories published on the first successful operation of the quantum cascade laser (QCL) in 1994 [27]. The QCL works fundamentally differently than the laser diode. Lasing transitions occur not from interband transitions, but between

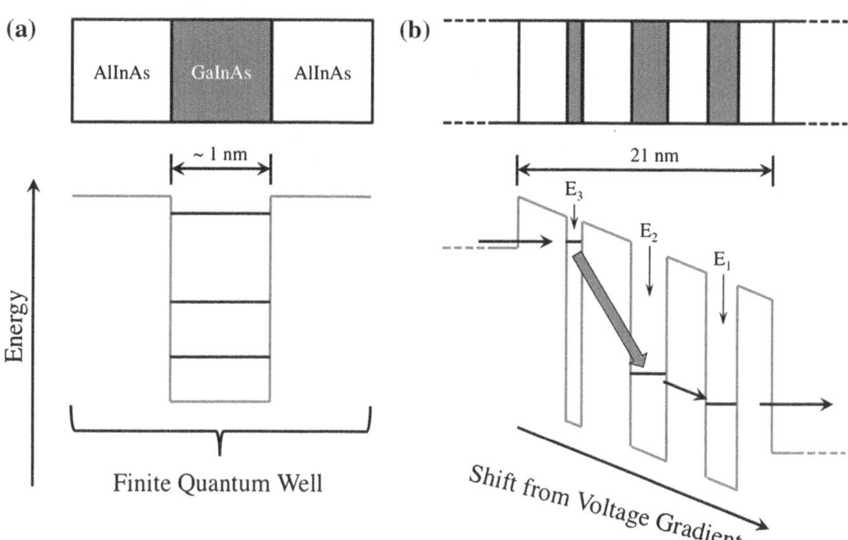

Fig. 2.20 Illustration of (**a**) quantum well schematic and corresponding energy diagram and (**b**) schematic of QCL heterostructure layers and corresponding energy diagram. Adapted from [27]

intersubband transitions within the conduction or valence band. Materials capable of QCL action can be manufactured to produce light in the mid-IR region (some even in the far-IR), a direct result of using intersubband transitions.

As illustrated in Fig. 2.20a, intersubbands arise from quantum wells created by using thin layers of *heterostructure* material. By stacking layers of various thicknesses of alternating AlInGa and GaInAs (seen in Fig. 2.20b), several intersubbands emerge that can be useful for lasing. The region shown in Fig. 2.20b can be repeated across the heterostructure multiple times, allowing for multiple sites of lasing action.

When a voltage is applied, the potential energy surface in the QCL is sloped downward. Electrons flow from the potential high point to the low point, and across the QCL active region. The energy level E_3 is rapidly populated from the electron flow, and its population is conserved due to a slow transition from E_3 to E_2. Following the $E_3 \rightarrow E_2$ transition, level E_2 is quickly depopulated by a rapid transition to E_1 (and even more rapid depopulation of E_1).

The barrier distances of heterostructure govern the kinetic mechanism behind QCL. A large barrier impedes the transition from $E_3 \rightarrow E_2$, as compared to smaller barriers between E_2 and E_1, and E_1 and the rest of the lattice. Intersubband transitions have much sharper bandwidths, in contrast to the broad bandwidths of diode lasers. Adjustments to the layer thicknesses of the heterostructure permit tuning of the output frequency. The example given in this chapter uses AlInAs (as a barrier) and GaInAs (as a well), but other materials have been used.

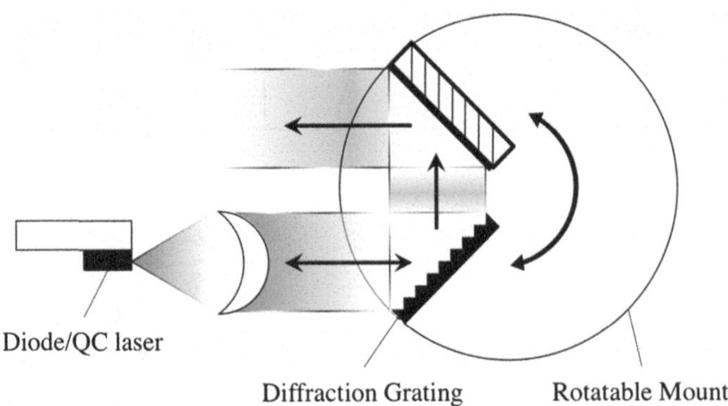

Fig. 2.21 Schematic of a Littrow configuration external cavity system. Adapted from [8]

In 2002, a report on a QCL heterostructure with dissimilar intersubband transitions, showed a broad output of frequencies ($\sim$6–8 μm) [28]. The tuning method of choice for QCL and diode lasers is an *external cavity* (EC) arrangement [8]. Figure 2.21 outlines the EC *Littrow configuration*, although other EC configurations exist. The basic principle of the EC is some form of wave-selective device (e.g., a diffraction grating) to reflect a narrow spectral width back into the cavity. In the above setup, the diffraction grating is mounted on a rotatable surface in order to scan through the frequency spectrum of the QCL. Optical setups for diode and QCL often use a lens to collimate the divergent beam and a temperature-controlling device. EC offers the widest tunability available for semiconductor lasers, with sharp spectral widths down to >0.001 cm^{-1}, although not in all cases.

2.7 Other Light Sources

A number of light sources that are or could be useful in ion spectroscopy or ion photodissociation experiments have not been discussed here. For instance, producing intense radiation at lower photon energies, such as in the far-IR (i.e., photons >20 μm), or, alternatively, very high energies, such as X-rays (i.e., nm radiation), requires a source like a free-electron laser (FEL). Without going into the details, a FEL makes use of an electron beam that is accelerated to relativistic speeds (i.e., MeV or even GeV) by a linear accelerator. When these electrons enter the oscillating magnetic fields of an undulator, they can emit a photon, whose frequency is determined by the kinetic energy of the electron and the magnetic field strength of the undulator. Operating a FEL requires a major capital expenditure, which is why they are typically operated as user facilities, where scientists can apply for "beam" time. There are merely a few of these user facilities in the world, limiting their availability to the wider scientific community.

Other light sources that have not been discussed here are gas discharge lamps, such as mercury lamps, deuterium lamps, given their very broad spectral profiles, constraining their photon irradiance at a particular wavelength is problematic.

2.8 Concluding Remarks

Considerable strides have been made in generating tunable, monochromatic light from the UV down to the mid-IR via benchtop light sources. Many of these light sources can be operated as turnkey benchtop systems and offer useful output characteristics, such as being widely tunable, having sharp bandwidths, and allowing pulsed timing. Nonlinear crystals based on OPO/A represent the state-of-the-art technology for experiments in combination with mass spectrometry. Further developments of semiconductor-based lasers, such as QCL, are likely to provide novel avenues in the future that have the potential to substantially reduce the cost and required expertise of operating these light sources.

References

1. Ionin AA (2007) In: Endo M, Walter RF (eds) Gas lasers. CRC Press, Boca Raton, Florida
2. Koechner W (2006) Solid-state laser engineering. Springer, Berlin
3. Vodopyanov KL, Levi O, Kuo PS et al (2004) Opt Lett 29:1912–1914
4. Woodworth JR, Rice JK (1978) J Chem Phys 69:2500
5. Schafer FP (ed) (1977) Dye laser, 2nd edn. Springer, Berlin Heidelberg, New York
6. Barker E, Adel A (1933) Phys Rev 44:185–187
7. Hooker SM, Webb CE (1994) Prog Quant Electron 18:227–274
8. Hugi A, Maulini R, Faist J (2010) Semicond Sci Tech 25:083001
9. Lioe H, O'Hair RA, Reid GE (2004) J Am Soc Mass Spectrom 15:65–76
10. Mino WK, Gulyuz K, Wang D et al (2011) J Phys Chem Lett 2:299–304
11. Lifshitz C (2002) Eur J Mass Spectrom 8:85
12. Siegman AE (1986) Lasers. University Science Books, Mill Valley
13. Shimoda K (1986) Introduction to laser physics, 2nd ed. Springer, Berlin
14. Wadt WR (1978) J Chem Phys 68:402
15. Freeman DE (1974) J Chem Phys 61:4880
16. Liu K, Li FQ, Xu HY et al (2013) Opt Commun 286:291–294
17. Zhou B, Kane TJ, Dixon GJ et al (1985) Opt Lett 10:62
18. Wagner GJ, Carrig TJ, Page RH et al (1999) Opt Lett 24:19–21
19. Adams JJ, Bibeau C, Page RH et al (1999) Opt Lett 24:1720
20. Welford D, Moulton PF (1988) Opt Lett 13:975–977
21. Brockman P, Bair CH, Barnes JC et al (1986) Opt Lett 11:712–714
22. Kildal H, Mikkelsen JC (1973) Opt Commun 9:315–318
23. Skauli T, Vodopyanov KL, Pinguet TJ et al (2002) Opt Lett 27:628–630
24. Bosenberg WR, Guyer DR (1993) J Opt Soc Am B 10:1716–1722
25. Peterson R, Feaver R, Powers P (2011) OSA technical digest (CD). NME7, Optical Society of America
26. Solutions P (2013) Photonic solutions: unit A, vol 2013 (40 Captains Road, Edinburgh, EH17 8QF)
27. Faist J, Capasso F, Sivco DL et al (1994) Science 264:553–556
28. Gmachl C, Sivco DL, Colombelli R et al (2002) Nature 415:883–887

Chapter 3
Ion Traps

Kerim Gulyuz and Nicolas C. Polfer

3.1 Introduction

Ion traps are devices that use electric and/or magnetic fields to spatially confine charged particles in vacuum for extended periods of time and they have evolved as an important tool in mass spectrometry since their first introduction in the 1950s. Although there are many designs that utilize electric and magnetic fields to manipulate ion motion, there are two types that are most commonly used. The *Penning trap*, first developed by Hans G. Dehmelt, uses a strong axial homogeneous magnetic field to confine the ions radially and an electrostatic field to restrict the ions' axial motion. The other commonly used ion trap, the *Paul trap*, developed by Wolfgang Paul and his colleagues employs oscillating radiofrequency electric fields in combination with a static DC field. Dehmelt and Paul were awarded the 1989 Nobel Prize in Physics for their contribution to the development of ion trap techniques.

Ion-trapping devices play a crucial role in laser spectroscopy experiments on ions. The spatial confinement of the ions in ion traps naturally facilitates overlap of the ion cloud with the focused beam of the light source. Moreover, the mass selective capabilities of ions traps provide for powerful manipulation so that ions of interest can be mass selected prior to photodissociation and that the photodissociation products can be mass detected. The fundamentals, designs, and operations of different ion-trapping devices useful for laser photodissociation are reviewed here. Their respective strengths and weaknesses with respect photodissociation measurements are discussed.

K. Gulyuz (✉) · N. C. Polfer
Department of Chemistry, University of Florida, Gainesville, FL 32611, USA
e-mail: kgulyuz@ufl.edu

N. C. Polfer
e-mail: polfer@chem.ufl.edu

N. C. Polfer and P. Dugourd (eds.), *Laser Photodissociation and Spectroscopy of Mass-separated Biomolecular Ions*, Lecture Notes in Chemistry 83,
DOI: 10.1007/978-3-319-01252-0_3,

3.2 Penning Traps

Penning traps, also widely known as the *Fourier transform ion cyclotron resonance* (*FT-ICR*) traps, allow excellent control of ions for many applications. The mass accuracy reaching sub-ppm and the ultrahigh resolving power makes FT-ICR devices suitable for experiments that require accurate mass measurements. The ultra-high vacuum (UHV) ($\sim 10^{-9}$ Torr) environment provides an almost collision-free environment for experiments that involve fast reactions. The possibility of extended trapping times from seconds to hours allows probing of slower reactions. Laser irradiation/photodissociation experiments are possible by a small modification in the setup to allow the laser light into the trap. Let us start with the basic theory of operation. For a more thorough review, interested readers may refer to review articles by Marshall [1–3], who vastly contributed to the field of FT-ICR mass spectrometry.

3.2.1 Basic Equations

In a Penning trap, the charged particle moves in a static uniform magnetic field and experiences a force according to the Lorentz force law.

$$\mathbf{F} = q\mathbf{v} \times \mathbf{B} \tag{3.1}$$

where q and **v** is the charge and velocity of the moving particle, respectively, and **B** is the magnetic field in which the particle is moving. Because of the cross-product in the equation, the direction of this force is perpendicular to the plane defined by **v** and **B**. As shown in Fig. 3.1, the path of an ion is bent into a circle if the ion has a velocity component that is perpendicular to the magnetic field.

If one defines the direction of the magnetic field **B** to be the along the negative z-direction and the magnitude to be equal to B_0, such that

$$|\boldsymbol{B}| = B_0 \tag{3.2}$$

then, in terms of the cyclotron motion, only the velocity component of the particle in the xy *plane*, v_{xy}, which we define as

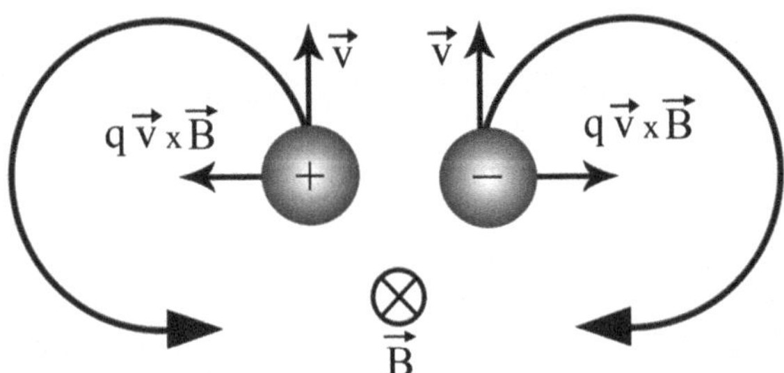

Fig. 3.1 Ion cyclotron motion for positively and negatively charged ions with velocities in the plane of paper, and the magnetic field pointing into the page

$$v_{xy} = \sqrt{v_x^2 + v_y^2} \tag{3.3}$$

is relevant. Knowing that the radial acceleration in the xy plane is

$$a_r = \frac{v_{xy}^2}{r} \tag{3.4}$$

one can rewrite Eq. (3.1) as

$$m\frac{v_{xy}^2}{r} = qv_{xy}B_0 \tag{3.5}$$

Since the angular velocity about the z-axis is defined by

$$\omega = \frac{v_{xy}}{r} \tag{3.6}$$

substituting this relation in Eq. (3.5) gives us, in the absence of any electric field, the unperturbed *ion cyclotron frequency* ω_c as

$$\omega_c = \frac{qB_0}{m} \tag{3.7}$$

One result that follows from this equation is that the ions of the same mass-to-charge ratio, *m/q*, have the same *cyclotron frequency*, but that their frequencies are independent of their velocities. As a numerical example, for a singly charged ion of 100 amu moving in a magnetic field of 4.7 T, the cyclotron frequency is about 4.5 MHz.

Figure 3.2 depicts the two most widely used geometries for Penning traps, which are also known as ion cyclotron resonance cells. The closed cell design comprises two trapping plates (T), as well as two excitation (E) and two detection (D) electrodes, which are positioned opposite to each other. The open cell design is composed of eight trapping electrodes, two excitation, and two detection plates.

Although the magnetic field confines the particles on the plane perpendicular to the field, the particles can still escape along the magnetic field lines. To prevent this, a restoring potential is required which has a potential minimum along the magnetic field axis. In a Penning trap, this is achieved by a DC electrostatic

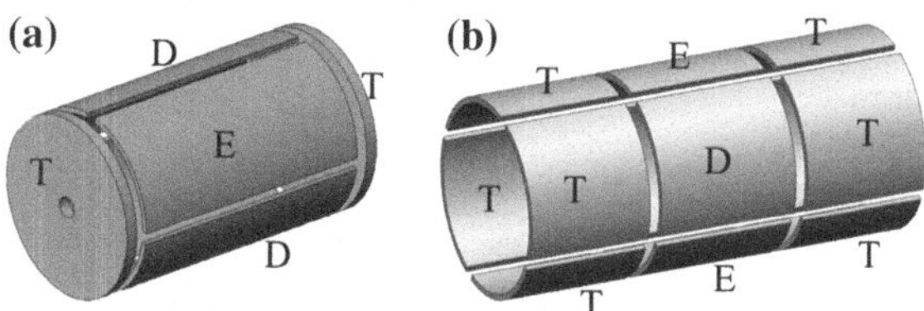

Fig. 3.2 Two of the most widely used Penning trap geometries; (**a**) closed cell and (**b**) open cell. The electrodes are labeled as *T*, *E*, and *D* representing the trapping, excitation, and detection electrodes, respectively

potential. The application of DC voltages on the trapping plates, as opposed to ground potential on the excitation and detection plates, results in a *quadrupolar potential* with a minimum in the center of the trap. For the Penning trap configuration depicted in Fig. 3.2b, the potential energy surface for the confinement in the axial direction is shown in Fig. 3.3. The combined effect of magnetic and electrostatic fields allows trapping of either positive or negative ions in three dimensions, depending on whether positive or negative voltages are employed on the trapping voltages.

The axial trapping results in axial harmonic oscillations with angular frequency

$$\omega_z = \sqrt{\frac{2qV_{\mathrm{trap}}\alpha}{md^2}} \tag{3.8}$$

where V_{trap} is the electrostatic voltage applied to the axial trapping electrodes, d is a measure of the trap size, and α is a constant that depends on the geometry of the trap. ω_z is typically of the order of 1–100 kHz. The motion in the radial plane is perturbed by the existence of this axial oscillation and two independent modes of oscillation are observed in the radial plane. One is called the *reduced cyclotron frequency* and is given by

$$\omega_+ = \frac{\omega_c}{2} + \sqrt{\frac{\omega_c^2}{4} - \frac{\omega_z^2}{2}} \tag{3.9}$$

and the other one is called the *magnetron frequency* and is given by

$$\omega_- = \frac{\omega_c}{2} - \sqrt{\frac{\omega_c^2}{4} - \frac{\omega_z^2}{2}} \tag{3.10}$$

In summary, there are three discrete ion motions in a Penning trap with angular frequencies ω_+, ω_-, and ω_z. For the motions that are in the same radial plane, a typical magnitude for the reduced cyclotron frequency ω_+ is 50–5,000 kHz, and 10–1,000 Hz for the magnetron frequency ω_-.

Other factors, like the space-charge of the ions, as well as imperfections in the trap geometry and magnetic field are typically accounted for by calibration. In other words, several ions of known mass-to-charge ratios (m/z's) are injected into the trap and their reduced cyclotron frequencies are measured and fitted. Then, the

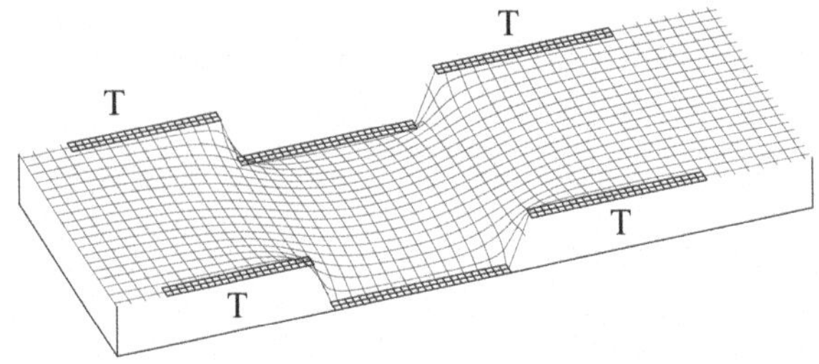

Fig. 3.3 The electrostatic potential energy surface of the Penning trap geometry depicted in Fig. 3.2b

m/z of an unknown ion can be found by measuring its reduced (or observed) cyclotron frequency.

3.2.2 Ion Excitation and Detection of the Signal

The cyclotron motion of the ions is detected by virtue of an *image current* that the charged particles induce on the detection electrodes. This requires a coherent ion packet that is moving in-phase at an appreciable orbit between the detection electrodes. Upon initial trapping, however, the ions have a random phase in their cyclotron motion; moreover, they are radially focused in the center by the magnetic field, and hence, their orbits are too small for detection. Figure 3.4a depicts the cross-section of a Penning trap, illustrating how the ions are accelerated to a larger detectable radius by applying an oscillating electric field on the excitation plates of the same frequency as the cyclotron frequency of the ion of interest. At any moment in time, opposite voltages $+V_0$ and $-V_0$ are applied to the oppositely placed *excitation plates* (E), resulting in a continuous increase in the cyclotron radius of the ion. This method is also called "*dipolar excitation.*" The resulting post-excitation ICR radius is typically on the order of 1 cm (and hence smaller than the radial dimensions of the trap) and is independent of the mass/charge ratio of the ions.

In order to detect the resulting excited ICR motion, another set of plates, called the *detection plates* (D), are utilized, as shown in Fig. 3.4b. As the packet of ions orbit, they generate a detectable induced current on the plates. The induced current has the same frequency as the ions' orbital frequency and its intensity is proportional to the number of charges. The number of charges in an ion packet in turn depends on the number of ions and the charge state of each ion. A range of ions of different m/z can all be excited to larger cyclotron radii by sweeping the excitation frequencies. The simultaneous motions of ions of different m/z's gives rise to a

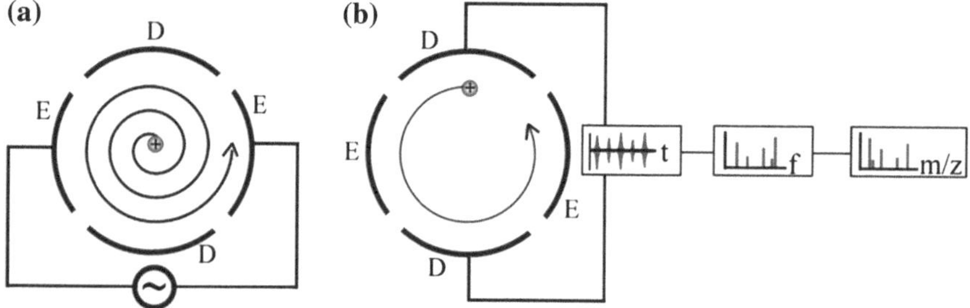

Fig. 3.4 (**a**) The ions are accelerated to a larger radius for detection by an oscillating voltage applied on the excitation plates. (**b**) The induced current on the detection plates is detected in the time domain; this signal is then decomposed to the frequency domain by a FT algorithm, followed by conversion to m/z values

complex image current on the detection plates. A Fourier transform (FT) analysis of the signal allows deconvolution into the various cyclotron frequency components, and thus (by calibration) the various *m/z* values.

3.3 RF Traps

In a *Paul trap* [4], no magnetic field is applied on the charged particles, and the three-dimensional confinement is achieved exclusively by using *radiofrequency* (rf) fields. Any oscillating force field provides the necessary restoring force to keep the ions in a certain trajectory. One of those types of electric fields has a form that depends linearly on the coordinates and thus has a uniform gradient; i.e., $\nabla \cdot \mathbf{E} =$ constant. One can express such an electric field in Cartesian coordinates as

$$\mathbf{E} = \mathrm{E}_0(\lambda x\,\mathbf{i} + \sigma y\,\mathbf{j} + \gamma z\,\mathbf{k}) \tag{3.11}$$

where $\mathbf{i}$, $\mathbf{j}$, $\mathbf{k}$ are the unit vectors, x, y, z are the coordinates, λ, σ, and γ are coefficients and E_0 is a position-independent coefficient, but which may be time-dependent. The motion can be analyzed separately in each coordinate, because there are no cross-terms, i.e., the electric field in the x-direction depends only on x and not on y or z. The *Laplace equation* in the absence of a charge is then reduced to

$$\nabla \cdot \mathbf{E} = 0 \tag{3.12}$$

which yields the relation

$$\boldsymbol{\lambda} + \boldsymbol{\sigma} + \gamma = 0 \tag{3.13}$$

One of the simplest solutions has $\lambda = -\sigma$ and $\gamma = 0$, which applies to the 2-dimensional quadrupole mass filter and linear ion trap case; another simple solution has $\lambda = \sigma$ and $\gamma = -2\sigma$, which applies to the three-dimensional quadrupole ion trap case. The electric field E can be expressed as a gradient of an electric potential Φ such that

$$\mathbf{E} = -\nabla\Phi \tag{3.14}$$

or in component notation along each axis as

$$E_x = -\frac{\partial\Phi}{\partial x}, \quad E_y = -\frac{\partial\Phi}{\partial y}, \quad E_z = -\frac{\partial\Phi}{\partial z} \tag{3.15}$$

Integration results in

$$\Phi = -\frac{E_0}{2}(\lambda x^2 + \sigma y^2 + \gamma z^2) \tag{3.16}$$

3.3.1 2D Quadrupole Ion Trap (Linear Ion Trap)

In the case where $\lambda = -\sigma$ and $\gamma = 0$, the problem reduces into two dimensions with the potential given by

$$\Phi = -\frac{E_0\lambda}{2}(x^2 - y^2) \tag{3.17}$$

A *quadrupolar potential* of this form can be established with a set of four electrodes with hyperbolic cross-sections as shown in Fig. 3.5, where say a positive potential is applied to two opposite electrodes. Conversely, a negative potential (of the same magnitude as the positive potential) is applied to the adjacent rods. If one defines the distance between two opposite electrodes to be $2r_0$, then the quadrupole potential can be written as

$$\Phi = \frac{\Phi_0}{r_0^2}(x^2 - y^2) \tag{3.18}$$

and this corresponds to *equipotential lines*, as shown in Fig. 3.5. For practical reasons, hyperbolic rods can be replaced by circular rods of radius r to a good approximation of the quadrupolar field by selecting $r = 1.148\ r_0$.

3.3.1.1 The Equations of Motion in a Quadrupolar Field

For an ion in an electric field, we have written earlier with the help of Newton's laws of motion that the force acting on the ion in an electric field is $e\,E$ (i.e., charge times electric field) which should be equal to $m\,a$ (i.e., mass times acceleration). As each component of the electric field only depends on its corresponding coordinate, one can separate the equation of motion in three coordinates, such that

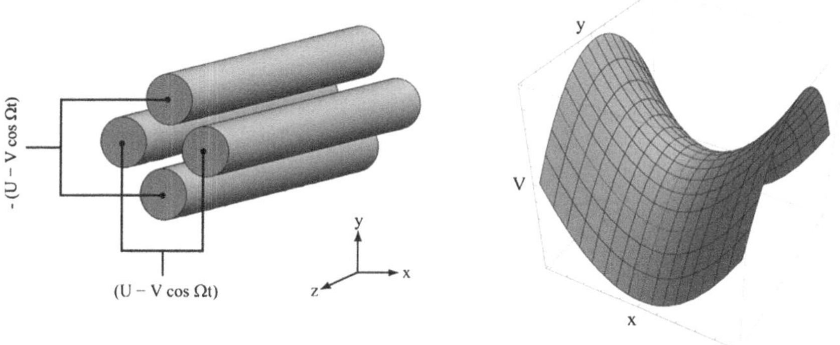

Fig. 3.5 Schematic of a linear ion trap with four rods (*left*) and the corresponding potential energy surface in the xy plane (*right*)

$$m\frac{d^2x}{dt^2} = eE_x, \quad m\frac{d^2y}{dt^2} = eE_y, \quad m\frac{d^2z}{dt^2} = eE_z \tag{3.19}$$

where E_x, E_y, and E_z are as defined in Eq. (3.15). Here, we choose the axial direction, on which the ions are introduced into the quadrupole mass filter, to be the z-direction. By inserting Eq. (3.18) into the equations above, we get

$$\begin{aligned} \frac{d^2x}{dt^2} + \frac{2e}{mr_0^2}\Phi_0 x = 0, \\ \frac{d^2y}{dt^2} - \frac{2e}{mr_0^2}\Phi_0 y = 0, \\ \frac{d^2z}{dt^2} = 0 \end{aligned} \tag{3.20}$$

Choosing Φ_0 to be time-independent gives us a *simple harmonic motion* in the xz plane and all the trajectories will have a finite amplitude and thus be stable in this plane. In the yz plane, however, the motion will be defocusing and unstable, making the ions deviate from the z-axis.

By choosing Φ_0 to be periodic in time, we can get an alternatively focusing and defocusing motion in both the xz and yz planes. However, not any random periodicity produces this kind of motion. The oscillation period of the time-varying potential must be short enough, such that the ion does not have enough time to escape from the quadrupole dimensions during the time interval when it was defocused. These constraints lead to defined stability conditions that must be satisfied for successful confinement.

Suppose we choose $\Phi_0 = U - V\cos\Omega t$, where U is a DC voltage applied to the electrodes and V is the zero-to-peak amplitude of the RF voltage oscillating with the angular frequency Ω. The equations of motion in x and y directions can then be written

$$\begin{aligned} \frac{d^2x}{dt^2} + \frac{2e}{mr_0^2}(U - V\cos\Omega t)x = 0, \\ \frac{d^2y}{dt^2} - \frac{2e}{mr_0^2}(U - V\cos\Omega t)x = 0 \end{aligned} \tag{3.21}$$

If we introduce the dimensionless parameters

$$\begin{aligned} a_u = a_x = -a_y = \frac{8eU}{m\Omega^2 r_0^2} \\ q_u = q_x = -q_y = \frac{4\,eV}{m\Omega^2 r_0^2} \\ \xi = \frac{\Omega t}{2} \end{aligned} \tag{3.22}$$

and make a change of variables, we get the canonical form of the *Mathieu equation*

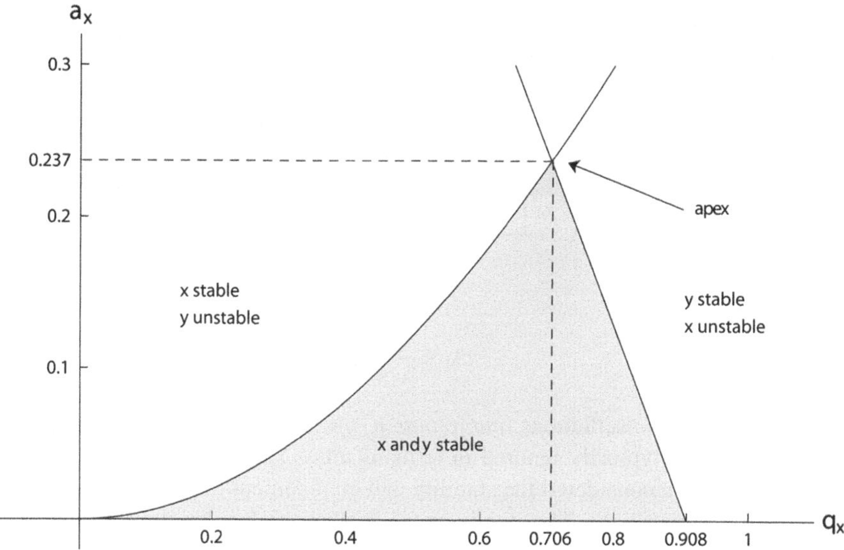

Fig. 3.6 The first region of stability for a 2D linear quadrupole ion trap

$$\frac{d^2u}{d\xi^2} + (a_u - 2q_u \cos 2\xi)u = 0 \tag{3.23}$$

There are certain values of a_x and q_x for which the motion is stable in the xz plane; i.e., the x value does not exceed r_0, thus forming a region of stability. A similar argument applies to a_y and q_y. As can be seen from the definitions of a_u and q_u, they depend on the form of the confinement potential, the mass of the ion of interest, and the quadrupole geometry. Thus, for a particular trap size and ion type, one can realize the proper voltages to be applied for a stable operation.

The first region of stability, shown in Fig. 3.6, where $0 < q < 0.908$ is particularly important because it contains the area with largest overlap of x-stable and y-stable regions. Consequently, this region is used in virtually all quadrupole mass filters. The advantage of representing the stability diagram in terms of these dimensionless parameters is that these a_u and q_u boundary conditions apply to every ion, independent of the ion's *m*/*z*, the trap dimensions, and the voltages that are applied.

3.3.1.2 Ion Motion in the 2D Trap and Capturing Ions in Three Dimensions

A detailed analysis of the stable solutions to the Mathieu equations (which we will not discuss here) yields motions of the ions in the trap that correspond to a superposition of some frequencies, the lowest of which is called the *macromotion* (or *secular*) *oscillation*. That frequency can approximately be written as

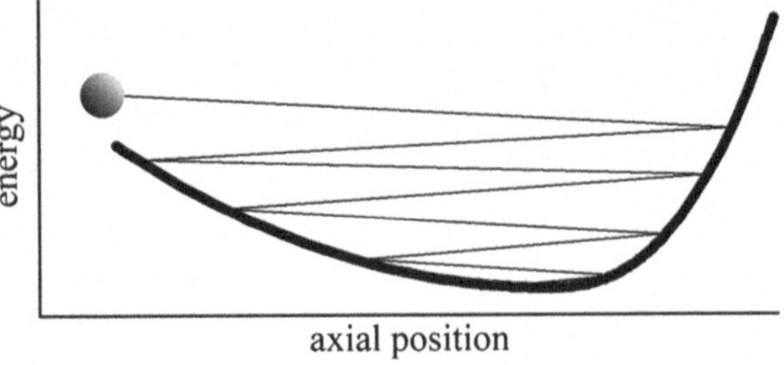

Fig. 3.7 The ion loses its kinetic energy via collisions as it travels through buffer gas until it is trapped at the *bottom* of the DC potential well created by the trap electrodes

$$\omega_0 \approx \frac{q\Omega}{\sqrt{8}} \tag{3.24}$$

The higher order oscillations (*micromotion*) are a direct result of the driving rf voltage; these are typically ignored in terms of mass spectrometric analysis.

So far, we have considered the stability and confinement/trapping of the ions in the x and y dimensions. The confinement in the axial z-direction requires an additional electric field with a potential minimum at the trap center. This condition can be achieved by adding two plates at the both ends of the quadrupole rod structure, in an arrangement similar to the Penning trap case shown in Fig. 3.2a. When loading the trap with ions, the potential barrier at the entrance of the trap is selected just below the kinetic energy of the incoming beam, so that the ions can enter the trap and the potential barrier at the exit plate is kept high, so that the ions bounce back into the trap. In the absence of any drag, the ions would exit the trap through the entrance plate. To prevent this, a *buffer gas* is introduced into the trap, and the ions are made to lose kinetic energy by collisions with the buffer gas (buffer gas cooling). As illustrated in Fig. 3.7, each collision reduces the kinetic energy of the ion, so that the ion cannot overcome the barrier(s) and is hence trapped.

A typical pressure range for the buffer gas is 10^{-6} to 10^{-3} Torr, leading to a sufficient rate of collisions. Over time, these collisions thermalize the ions; in other words, the kinetic energy distribution of the ions approaches the temperature of the collision gas. A lowering of the kinetic energy causes a reduction in the size of the ion cloud to the bottom of the trap well. However, this shrinking is counteracted by the Coulomb repulsion force between the ions. High densities of ionic charges, commonly referred to as "*space charge*" effects, present an upper limit to the number of ions that can be stored in ion traps. Direct information on the shapes of the ion clouds can be obtained by, for example, *laser tomography*; by scanning a laser beam along the diameter of the ion cloud, the density profile can be imaged.

While the buffer gas is essential to allow efficient trapping and to "cool" the ions to a compact ion cloud, the presence of the buffer gas can present problems for mass detection and for efficient laser photodissociation. For this reason, in many laser spectroscopy experiments, the buffer gas is introduced in a pulsed fashion, so that the pressure spikes at some times in the experiment, followed by pump-down to lower pressures. The higher pressure regime favors ion trapping, while the lower pressures minimize ion cooling (and other adverse effects) during laser irradiation.

3.3.1.3 Mass Selectivity as Linear Mass Filter

It is possible to use quadrupole rods as a mass filter that selectively transmits ions with a certain *m*/*z* range, while blocking the other ions. When all four rods are floated at the same DC voltage, there is no DC field between the rods (i.e., $U = 0$), and consequently, the quadrupole is said to be operated on q_u axis (i.e., $a_u = 0$). In this *RF-only mode* of operation, all ions above a certain *m*/*z* value are transmitted, reflecting its function as a high-pass filter.

In *mass-selective transmission* mode, opposite polarity DC voltages are applied to both pairs of rods, in addition to the RF voltages that are applied. For one pair of rods, say the one on the xz plane, a voltage of U–V cosΩt is applied at any moment in time. Due to the positive potential *U*, an electrostatic minimum exists between the two rods for positive ions, causing a focusing effect for positive ions. Nonetheless, the oscillating RF potential can defocus lighter ions, resulting in increasing amplitudes, until they are lost either by ejection from the rod assembly or by crushing into the rods. The heavier (positive) ions on this plane are affected less and maintain stable trajectory. In other words, the field on the xz plane acts as a high-pass filter.

On the yz plane, however, where the pair of rods has a voltage of −(U–V cosΩt) applied, both light and heavy cations are defocused by the negative DC potential of the rods. The RF field can refocus the light ions toward the *z*-axis more easily than the heavier ions, effectively making the field on the yz plane a low pass filter.

Thus, by a clever choice of the DC to RF ratio, one can obtain a transmission of ions in a narrow *m*/*z* range. The entire mass spectrum can be scanned by ramping the DC and RF potentials at a fixed U/V ratio, as shown in Fig. 3.8. Note that anions respond in the opposite fashion to cations, and hence, the focusing and defocusing effects in the xz and yz planes, respectively, are inverted.

Another view to this picture is to consider the stability diagram in U-V space, in which ions with different *m*/*z* ratio will have different regions of stability, as shown in Fig. 3.9.

Here, the *operating line* or the *scan line* intersects the stability regions of various ions near their *apexes*, allowing the transmission of only one particular *m*/*z* for those particular values of *U* and *V*. A mass spectrum is obtained by detecting ions as a function of the U/V scan line, which is operated as a function of time. The steepest scan line, approaching a maximum slope of 0.336, exactly intersects the apex of each stability diagram and hence provides the highest separation and therefore best resolution between different *m*/*z* values. On the other hand, a steeper scan line provides little overlap with the various stability regions and hence comes at a cost of lower ion transmission. The trade-off between mass resolution and ion

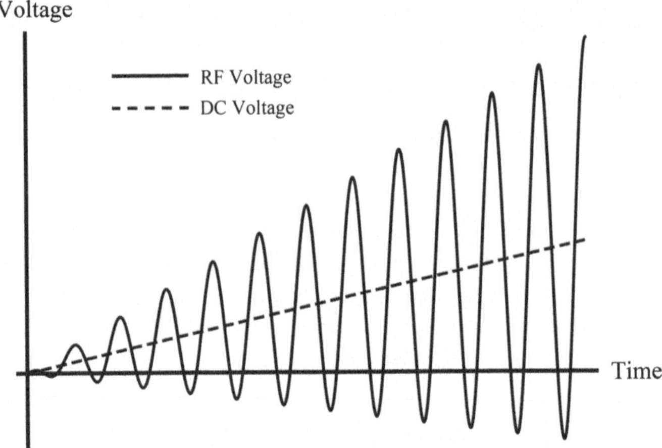

Fig. 3.8 The DC and RF voltages are scanned at a fixed U/V ratio to only transmit ions in a narrow *m/z* range at a given moment in time

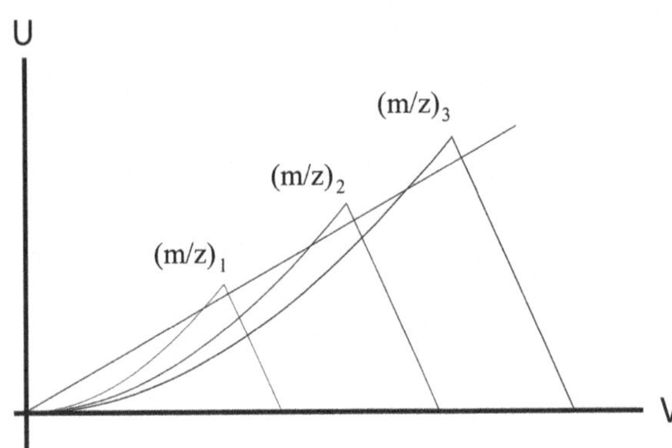

Fig. 3.9 Stability regions as a function of *U* and *V* for different *m/z* ions, where $(m/z)_1 < (m/z)_2 < (m/z)_3$. The ion *m/z* for which the scan line intersects its stability region at a particular *U* and *V* values is transmitted

signal intensity, and hence sensitivity, is dependent on the slope of the scan line, where flatter scan lines result in lower resolution, but higher ion signal intensities.

3.3.2 3D Quadrupole Ion Trap

Another simple solution to Eq. 3.13 is the case where $\lambda = \sigma$, $\gamma = -2\sigma$. In that situation, the electric potential can be expressed by

$$\Phi = -\frac{E_0\lambda}{2}(x^2 + y^2 - 2z^2) \tag{3.25}$$

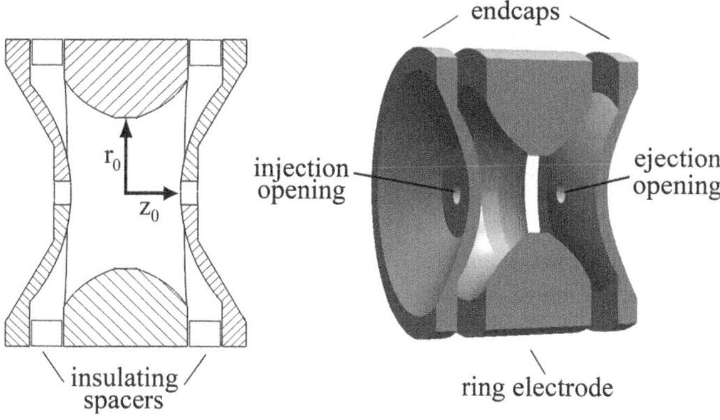

Fig. 3.10 Schematics of a 3D quadrupole ion trap

Such a form of the electric potential can be created by the three electrodes in a 3D quadrupole ion trap, a center *ring electrode* and two *endcap electrodes*, all with hyperbolic surfaces, as shown in Fig. 3.10. The inner radius of the ring electrode is defined as r_0, and the shortest distance between the endcaps is given as $2z_0$. Externally generated ions enter the 3D ion trap through apertures in the endcap electrodes. The detection scheme also typically involves ejection of the ions via an opening in the endcap electrodes. Due to ion optics constraints, the injection and ejection openings are generally on opposite sides.

3.3.2.1 The Equations of Motion in a 3D QIT

In ion-trapping mode, DC potentials are applied to the endcaps (i.e., usually ground potential), while DC and RF potentials are applied to the ring electrode, yielding an overall time-dependent potential of $\Phi_0 = U\text{–}V\cos\Omega t$ (where U = DC potential difference between the endcap and ring electrodes). Like in the case of the linear ion trap, the ion motion can be investigated separately in different directions, in this case the radial (r) and the axial (z) directions. Expressing $x^2 + y^2$ as equal to r^2, the equation of motion on the radial direction can be written as

$$\frac{d^2r}{dt^2} + \frac{2e}{m(r_0^2 + 2z_0^2)}(U - V\cos\Omega t)r = 0 \tag{3.26}$$

And in the axial direction we have

$$\frac{d^2z}{dt^2} + \frac{4e}{m(r_0^2 + 2z_0^2)}(U - V\cos\Omega t)z = 0 \tag{3.27}$$

Once again, the general form of the *Mathieu equation* 3.23 can be obtained by making the following substitution to dimensionless parameters.

$$a_u = a_z = -2a_r = \frac{-16\text{eU}}{\text{m}\Omega^2(r_0^2 + 2z_0^2)}$$
$$q_u = q_z = -2q_r = \frac{8\text{eU}}{\text{m}\Omega^2(r_0^2 + 2z_0^2)} \tag{3.28}$$
$$\xi = \frac{\Omega t}{2}$$

The stability conditions dictate that the position coordinate of an ion must never reach r_0 or z_0. The complete solution to the Mathieu equation is non-trivial, but for reasons of convenience a new parameter, β_u, is introduced, which can approximately be written as

$$\beta_u \approx \sqrt{a_u + \frac{q_u^2}{2}} \tag{3.29}$$

The stable ion trajectories require $0 < \beta_u < 1$. Once again the first stability region has the most practical importance. In Fig. 3.11, only the a_z versus q_z graph is shown, instead of a_u versus q_u, to avoid confusion. Note that the a_z and q_z parameters refer to the stability criteria with respect to the z-direction. In other words, a_z and q_z values outside of this stability diagram would result in a loss of the ions along the z-direction. It can be seen that the $\beta_z = 1$ line intersects with the q_z axis at $q_z = 0.908$. As will be discussed below, 3D quadrupole ion traps are generally operated along the q_z axis (i.e., $a_z = 0$ and hence $U = 0$), meaning that no DC potential difference is applied between the encaps and the ring and therefore giving special significance to the $q_z = 0.908$ value at the stability limit.

The application of the RF voltage on the ring electrode creates an effective potential well in the trap called the *Dehmelt pseudopotential well* [5], which can be written in the z-direction as

$$D_z = \frac{4\text{eV}^2 z_0^2}{\text{m}\Omega^2(r_0^2 + 2z_0^2)^2} = q_z \frac{V}{8} \tag{3.30}$$

where in the last part of the equation we have assumed $r_0^2 = 2z_0^2$ and inserted q_z using Eq. 3.28.

The space-charge effect is again a limiting factor for the number of ions that can be trapped simultaneously as the space-charge potential of a trapped ion cloud becomes comparable with the pseudopotential. For example, for an ion trap size of $r = 1$ cm, operating at $V = 1{,}000$ V at $\Omega = 2\pi * 1$ MHz, the maximum ion density for m/z 500 ions is 8.1×10^7 cm^{-3}.

3.3.2.2 Quadrupole Ion Trap as Mass Analyzer

The 3D quadrupole ion trap can be used as a mass analyzer after ions of different masses are trapped together. There are two main methods that can be used in mass

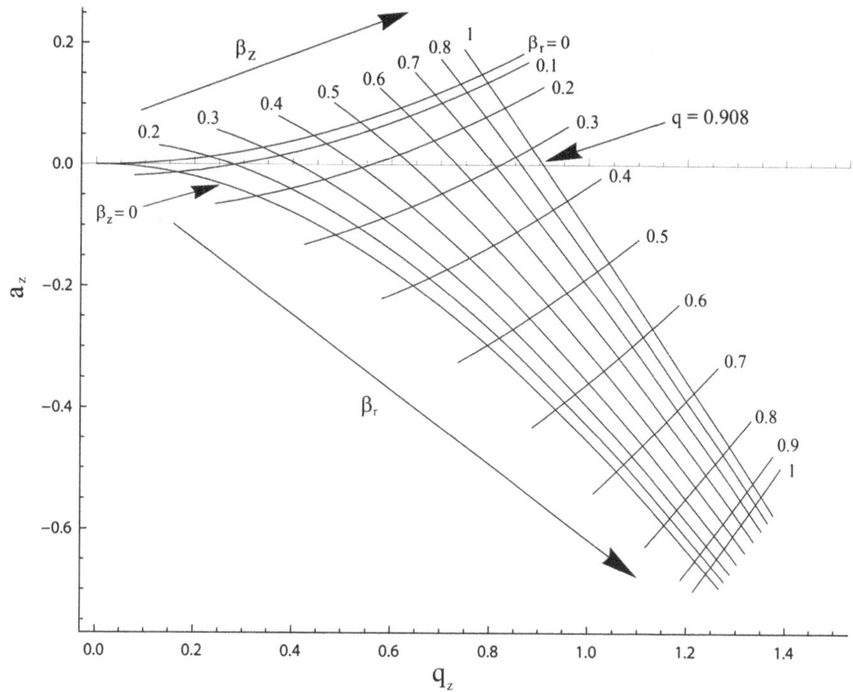

Fig. 3.11 The stability diagram for a 3D quadrupole ion trap. The stability region is confined by the curves where $0 < \beta_u < 1$

analysis. One method relies on the ejection of the ions at the stability limit and the other uses the secular frequency of the ions for resonant ejection.

Ion Ejection at the Stability Limit

When the two endcaps and the ring electrode are floated at the same DC potential (i.e., $a_z = 0$), the stability of an ion in the 3D trap exclusively depends on its q_z value, and thus, the RF frequency Ω and the RF amplitude V applied to the ring electrode. For given values of Ω and V, q_z is inversely dependent on the ion's m/z value (i.e., m/e, see Eq. 3.28). As shown in Fig. 3.12a, larger m/z ions are further from the ejection limit at $q_z = 0.908$ than lighter ions. As the amplitude of the RF voltage is increased, the q_z value for a particular ion reaches the stability limit and is consequently ejected from the trap in the z-direction (Fig. 3.12b). A mass spectrum is obtained by linearly increasing the RF voltage V, causing ions of progressively larger m/z values to exit the 3D ion trap and impact on an ion-selective detector (Fig. 3.12c). Note that ion ejection along the stability limit could in principle also be achieved by changing the RF frequency Ω; however, technically this is much more challenging. A drawback of the stability limit approach is that high m/z ions require impractically large RF voltages, imposing constraints on power supplies and potentially leading to electrical discharges.

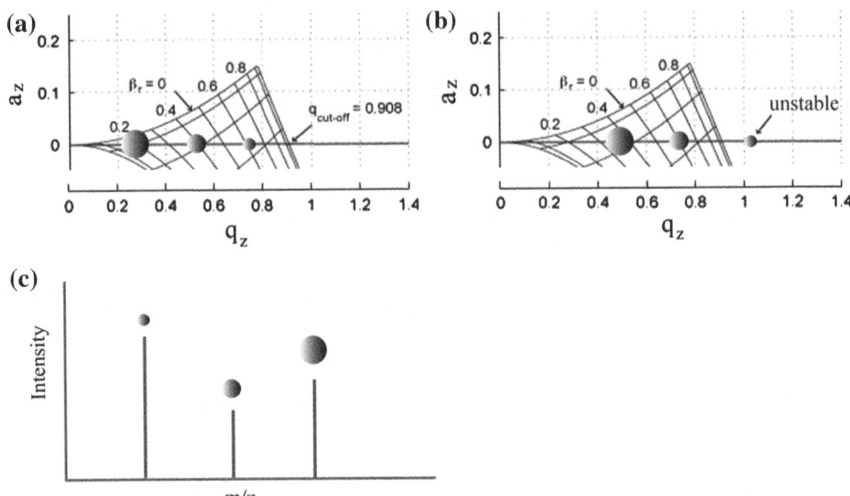

Fig. 3.12 Ejection of ions from the trap by raising the RF voltage amplitude, *V*. **a** The ions are stable in the trap. **b** When *V* is increased, the ions will move to higher q-values, causing the low-mass ions to be ejected from the trap first. **c** When *V* increased further, the higher-mass ions are also ejected, an effect which can be recorded with an ion-selective detector

Resonant Ejection

An alternative approach to eject ions from a 3D QIT is to resonantly excite their periodic motions in the trap. The trapped ions in a QIT oscillate under the parabolic potential wells in the axial and radial directions with so-called secular frequencies ω_z and ω_r, respectively, where

$$\omega = (n + \frac{2\beta}{2})\Omega \tag{3.31}$$

For the lowest *secular frequency* ($n = 0$) in the axial direction, we can write

$$\omega_z = \frac{\beta_z \Omega}{2} \tag{3.32}$$

The *secular frequency* of an ion depends on its *m*/*z* value, the RF frequency Ω and amplitude *V* applied to the ring electrode, as well as the trap dimensions. Table 3.1 gives numerical values for q_z, β_z, and ω_z values for a series of singly charged ions trapped in a given 3D quadrupolar ion trap. The dimensions of the trap are assumed to be $r_0 = 10$ mm, $z_0 = 7$ mm. The RF voltage is taken to be $V = 1{,}000$ V with $\Omega = 2\pi{*}1$ MHz, and no DC voltage. The lowest *m*/*z* ions that can be trapped under the above conditions is 109. This is also known as the low-mass cutoff, as no ions below *m*/*z* 109 can be trapped. In order to enhance the amplitude of the secular ion motion, an RF voltage resonant with the secular frequency ω_z is applied to the endcaps. When this amplitude becomes sufficiently large, the ions will no longer be stable and they will be ejected in the *z*-direction

Table 3.1 Numerical values for q_z, β_z, and ω_z for some selected ions with different masses

m (amu)	q_z	β_z	ω_z (kHz)
100	0.986	No real solution	No stable motion
300	0.329	0.238	748
500	0.197	0.140	440
1,000	0.099	0.070	220
2,000	0.049	0.035	110
5,000	0.020	0.014	44

The dimensions of the trap are assumed to be $r_0 = 10$ mm, $z_0 = 7$ mm. The RF voltage is taken to be $V = 1{,}000$ V with $\Omega = 2\pi * 1$ MHz, and no DC voltage

and will strike a detector. Resonant ejection using the secular motion of the ions is the method of choice for mass isolations in an MS/MS experiment, as ions of both larger and smaller *m/z* values than the ion of interest can be ejected from the trap. Conversely, ion ejection at the stability limit only allows ejection of ions below a particular m/z value (which happens to have a critical $q_z = 0.908$).

3.3.3 Multipole Devices

There are a multitude of designs of ion storage and ion transmission devices. A group of those devices, generally referred to as *multipoles*, utilize an even number of parallel rods that are symmetrically placed to form a tube along the axis on which the ions are transmitted. A *hexapole*, for instance, is composed of six rods, placed at an angle of 360°/6 = 60° to adjacent rods. A hexapole is operated as an ion focusing device (in the radial direction) by applying an in-phase RF voltage on a set of three non-adjacent rods, while applying the corresponding out-of-phase RF voltage on the remaining three rods. Along with quadrupoles and hexapoles, *octopoles* (i.e., eight rods) are commonly used in mass spectrometric instrumentation. Larger multipoles, involving more than eight rods, are typically confined to custom instrumentation.

The effective potential that an ion experiences while in a multipole can be expressed as

$$U(r) = \frac{n^2 e^2 V^2}{4m\Omega^2 r_0^2} \left(\frac{r}{r_0}\right)^{2n-2} \tag{3.33}$$

where $2n$ is the total number of rods and r is the distance from the central axis between the rods. This means that $U(r)$ varies as $(r/r_0)^2$ for a quadrupole, as $(r/r_0)^4$ for an hexapole, as $(r/r_0)^6$ for an octopole, etc. Because $(r/r_0) < 1$, this effectively translates into a flatter focusing potential for a hexapole compared with a quadrupole when an ion moves away from the central axis. In other words, a quadrupole has the highest restoring force, resulting in an increased transmission of a selected ion. However, because of the n^2 dependence of $U(r)$, a quadrupole has the lowest overall amplitude of the potential well compared with higher multipoles. A direct consequence of this relationship is the narrower mass range that can be transmitted by quadrupoles. Therefore, for a simultaneous transmission of a wide range of ions, higher order multipole devices are preferred, even though they are not as effective as a quadrupole in focusing the ions. To approximate the curvature of the multipole potential in Eq. 3.33, the radius R of the rods should satisfy [6]

$$R = \frac{r_0}{n - 1} \tag{3.34}$$

For example, if a trap diameter of $2r_0 = 10$ mm is desired with $2R = 1$ mm diameter rods, then the number of rods, $2n$, should be 22. In fact, 22-pole traps are employed in several experimental setups [7–9] involved in cryogenic laser spectroscopy experiments. It should also be noted that, unlike quadrupoles, the mass-selective properties of multipoles are very limited, merely allowing lower mass ions to be pushed beyond the stability limit.

3.4 Cold Traps

To study the ions at temperatures close to absolute zero, where most of them are in their ground vibrational state, ions are stored in *cryogenic traps*. The thermalization of molecular ions with a buffer gas in a 22-pole trap was shown to provide an ensemble of cold ions for various experiments [6]. Applications of such traps in IR spectroscopy will be discussed in Chap. 4. Here, we will give an introduction on the structure and design properties of cold traps.

To create an ensemble of cold ions, the ions must be stored in an environment that has minimum thermal contact with the rest of the instrument, which is at room temperature. Ions are insulated from *blackbody radiation* by enclosing the cryogenic trap with a heat shield operated at a low temperature (e.g. 50 K). Moreover, good vacuum conditions prevent any collisional heating of the ions. In order to attain ultra-cold temperatures (i.e., $<$10 K), higher order multipole traps are more suitable, as the flatter trapping potentials reduce RF heating. The RF heating in a 3D quadrupole ion trap makes this more challenging, particularly if the trap is operated at higher q-values ($q > 0.3$). Nonetheless, cryogenic 3D quadrupole ion traps have been successfully used to cool ions down to slightly elevated cryogenic temperatures (i.e., 12–30 K), which is sufficiently cold to tag van der Waals-bound atoms or molecules [10].

The implementation of a cryogenic 3D quadrupole ion trap is depicted in Fig. 3.13. The trap is surrounded by a heat shield, usually made from stainless steel or copper, with polished surfaces to increase reflectivity and thus reduce the heat load caused by thermal radiation. The shield has apertures on both sides to allow ions to enter and exit the trap. The trap itself is mounted on a cold head of a helium-cycle cryostat through a copper block (red), separated by a sapphire (blue) insulating layer. The sapphire material acts as an electrical insulator but also

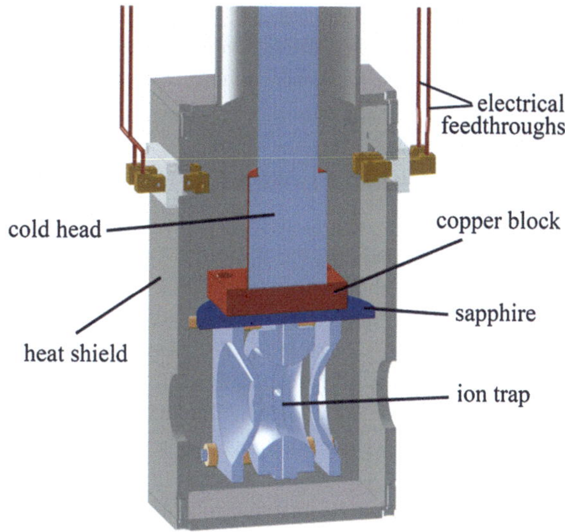

Fig. 3.13 Schematics of a cryogenic 3D quadrupole ion trap

provides adequate thermal conductance to cool the electrodes to the temperature of the cryostat cold head. The design of a cryogenic 3D ion trap is made simpler by the fact that the physical and thermal connections to all three electrodes (i.e., endcaps and ring electrode) are achieved via the same contact point at the top. Electrical connections (not shown) are made to each electrode individually via feedthroughs through the heat shield. Gas is introduced into the heat shield via a pulsed valve (not shown), which can be mounted on the bottom side of the heat shield. Care must be taken that the thermal load from the electrical connections and the pulsed gas do not significantly elevate the temperature of the electrodes with respect to the cold head.

Optical access to the trap center is possible via holes drilled into the ring electrode. This hole should be small to reduce the blackbody radiation generated by room temperature surfaces but should also be large enough to accommodate the laser beam. To reduce the beam waist of the laser beam, a focusing lens is usually employed.

3.5 Comparison of Traps

Many factors come into play when determining which trap is best suited for which type of experiment. Here, we will focus on the factors of these setups with regard to their usefulness in photodissociation experiments. Some of these factors include:

- the simplicity of construction and compactness of the design,
- mass accuracy and mass resolution,

- mass isolation capabilities,
- signal-to-noise ratio of the detector,
- trapping volume and trap capacity,
- (maximum) storage time,
- laser overlap with the ion cloud,
- and others, such as RF heating, effects of buffer gas, etc.

Some of those factors may also be related to each other. For example, laser overlap also depends on the trapping volume, which in turn determines the extent of RF heating in the trap.

FT-ICR instruments are known for their excellent mass accuracy (i.e., subppm) and ultra-high mass resolution. The trapping times that extend from subsecond to hours and the ultra-high vacuum environment allow for both fast and slow reactions. Nonetheless, an indispensible part of the FT-ICR setup is the superconducting magnet, which occupies large volumes, requires cryogenic maintenance and comes at a substantial price tag. Important factors in photodissociation experiments are ease of introduction of the laser beam into the mass spectrometer, as well as overlap of the laser beam with the ion cloud. The introduction of the laser beam into an ICR cell is usually directed on the central axis of the ICR cell, as shown in Fig. 3.14a, also comparing this approach to other ion trap setups. The ion cloud has an ellipsoid shape that extends along the axis of the ICR cell. This

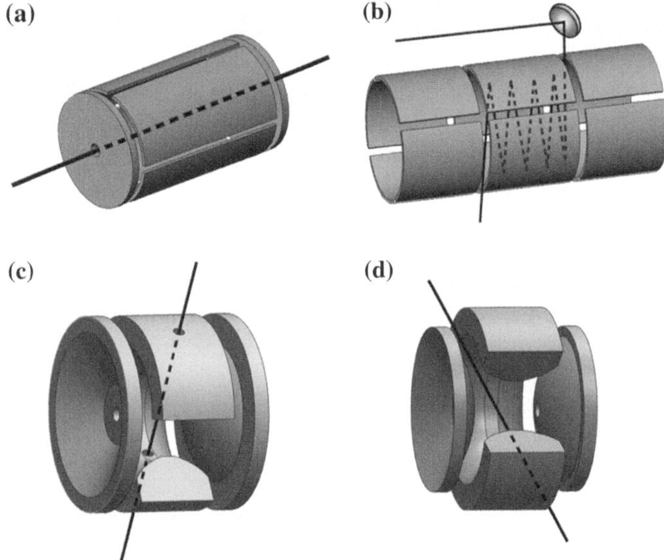

Fig. 3.14 Some methods for introduction of the laser in trapping volume. For an ICR cell, (**a**) the laser can be introduced on the central axis or (**b**) it intersects the ion cloud at several locations in a multipass arrangement. For a quadrupole ion trap, (**c**) the laser can be focused on the ion cloud through apertures drilled on the ring electrode or (**d**) along the common asymptote of the endcap and ring electrodes

elongated shape makes it difficult to provide a high photon flux for the majority of the ions, as the focal point of the laser beam is much smaller than the size of the ion cloud. In multi-pass arrangements (Fig. 3.14b), the laser beam can be focused onto the ion cloud at several locations, thus maximizing photon flux and providing enhanced overlap. In one of these approaches, the reflective surfaces of the detection plates were employed to re-focus the laser beam onto the ion cloud [11]. Cryogenic FT-ICR traps have already been employed in a number of ion spectroscopy experiments [12].

The 2D linear ion trap (LIT) is the most sensitive trap for ion photodissociation experiments, due to the high trapping efficiency and the large trapping capacity, combined with ultra-sensitive detection. In contrast to the less sensitive induced-current detection scheme in FT-ICR, the ions are destructively impacted onto a multiplier. In contrast to 3D quadrupolar ion traps, there is no quadrupolar field along the central axis, considerably enhancing the trapping efficiency for externally injected ions. Because of the necessity of a buffer gas to facilitate ion trapping, the LIT is usually operated at higher pressures (i.e., mTorr). However, the existence of high pressure raises a challenge for photodissociation experiments. To overcome this challenge, the buffer gas may be introduced in pulses via a pulsed valve, synchronized with the arrival of the ion beam to be trapped. Once the ions are trapped, the buffer gas is allowed to be pumped away, and the laser beam is introduced after the base pressure is attained. Similar to FT-ICR setups, the shape of the ion cloud is ellipsoidal along the central axis, and the laser beam is usually introduced on the central axis. A drawback of the LIT vis-à-vis FT-ICR instruments is its limited mass resolution, which typically only allows unit mass separation. No cryogenic LITs have so far been implemented in ion spectroscopy measurements, despite their great promise.

The 3D quadrupole ion trap (QIT) is not as effective as a LIT in trapping efficiency and trap capacity and hence suffers from lower sensitivity. However, a major advantage of a 3D QIT is the fact that the ion cloud is nearly spherical, and therefore, it is possible to intersect a focused laser beam with the majority of the ion cloud. The laser beam can be introduced into a QIT via the apertures drilled on the ring electrode, as depicted in Fig. 3.14c. The effect of this modification is small on the quadrupolar field and the trapping efficiency, and thus, it is the preferred method of modification due to its ease in implementation. It is also possible to introduce the laser beam without disturbing the hyperbolic surfaces, i.e., along the common asymptote of the endcap and ring electrodes (Fig. 3.14d). The use of cryogenic QITs has already been demonstrated in ion spectroscopy experiments.

Multipole traps are highly suitable for cold spectroscopy experiments because of the flatter focusing potential that reduces the RF heating effects. On the other hand, they are more limited as mass-selective devices and only useful for ejecting low-mass ions at the stability limit.

In summary, a wealth of ion-trapping devices is available to carry out ion photodissociation in combination with light sources. Their relative benefits and drawbacks depend on the types of experiments that are envisaged. Hybrid mass spectrometers combine multiple mass spectrometric devices and can thus overcome weaknesses of some components. For instance, 3D quadrupole ion traps have been coupled to time-of-flight (ToF) analyzers for higher resolution and faster (i.e., μs) detection [13, 14]. Linear ion traps have been coupled to both FT-ICR [15] and orbitrap mass analyzers [16] to increase the mass resolution and mass accuracy of detection. The development of cryogenic ion traps still remains in its infancy, and no commercial mass spectrometry manufacturer produces such instruments at this time, despite their promise in cold spectroscopy experiments.

References

1. Marshall AG (1996) Int J Mass Spectrom Ion Proc 137/138:410
2. Marshall AG, Hendrickson CL, Jackson GS (1998) Mass Spec Rev 17:1–35
3. Marshall AG (2000) Int J Mass Spectrom 200:331–356
4. Paul W, Steinwedel H (1953) Z Naturforsch 8A:448–450
5. Major FG, Dehmelt HG (1968) Phys Rev 170:91–107
6. Gerlich D (1992) Adv Chem Phys 82:1–176
7. Schlemmer S, Lescop E, von Richthofen J et al (2002) J Chem Phys 117:2068–2075
8. Rizzo TR, Stearns JA, Boyarkin OV (2009) Int Rev Phys Chem 28:481–515
9. Wang YS, Tsai CH, Lee YT et al (2003) J Phys Chem A 107:4217–4225
10. Leavitt CM, Wolk AB, Fournier JA et al (2012) J Phys Chem Lett 3:1099–1105
11. Polfer NC, Oomens J (2007) Phys Chem Chem Phys 9:3804–3817
12. Wong RL, Paech K, Williams ER (2004) Int J Mass Spectrom 232:59–66
13. Michael S, Chien M, Lubman D (1992) Rev Sci Instrum 63:4277–4284
14. Mano N, Kamota M, Inohana Y, Yamaguchi S, Goto J (2006) Anal Bioanal Chem 386:682–688
15. Syka JEP, Marto JA, Bai DL et al (2004) J Proteome Res 3:621–626
16. Hu Q, Noll RJ, Li H et al (2005) J Mass Spectrom 40:430–443

Chapter 4
Infrared Photodissociation of Biomolecular Ions

Nicolas C. Polfer and Corey N. Stedwell

4.1 Infrared Photodissociation Spectra of Mass-separated Biomolecular Ions

The analytical power of mass spectrometry (MS) is based on

- Its ability to separate thousands of analytes in complex mixtures by virtue of mass (-to-charge) and
- Its ultrasmall sample requirements

Despite these inherent strengths, the structural information from mass measurements alone is often insufficient. Further, structural information comes from ion dissociation, for instance by increasing the internal energy of the ion by energetic collisions. The focus of this chapter is on obtaining enhanced structural information on biomolecules by recording infrared (IR) spectra on these ions.

The grand strategy for employing vibrational spectroscopy to mass-separated biomolecular ions in a bioanalytical scheme is depicted in Fig. 4.1. Liquid chromatography (LC) is often employed to separate complex mixtures prior to analysis by MS. In IR photodissociation spectroscopy, the IR spectra of mass-separated analytes are recorded. In principle, the vibrational spectrum of an ion provides information on

- The presence/absence of particular chemical groups
- The location of the charge carrier
- The symmetry of the ion
- Its secondary structure
- Its hydrogen-bonding interactions

N. C. Polfer (✉) · C. N. Stedwell
Department of Chemistry, University of Florida, Gainesville, FL 32611, USA
e-mail: polfer@chem.ufl.edu

C. N. Stedwell
e-mail: stedwell@chem.ufl.edu

N. C. Polfer and P. Dugourd (eds.), *Laser Photodissociation and Spectroscopy of Mass-separated Biomolecular Ions*, Lecture Notes in Chemistry 83,
DOI: 10.1007/978-3-319-01252-0_4,

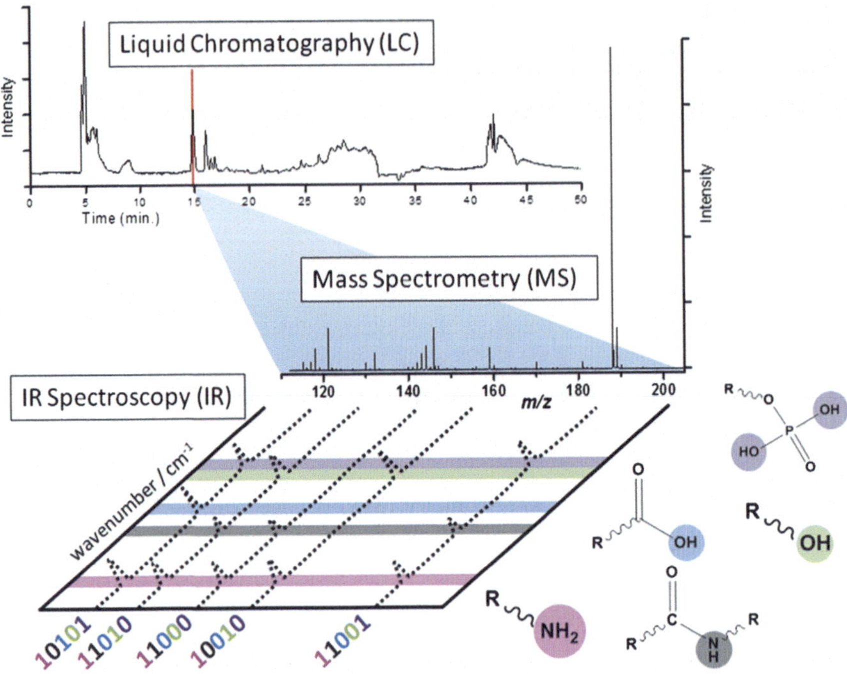

Fig. 4.1 Schematic for liquid chromatography separation, mass spectrometry measurement, and infrared spectroscopy measurement of mass-separated ions

Figure 4.1 emphasizes the compositional information from IR spectra, based on diagnostic IR band positions. This IR information can be thought of in a binary format, where various diagnostic bands are either present (indicated by "1") or absent (indicated by "0"). Recording IR spectra of ions requires tunable IR lasers (introduced in Chap. 2). The present chapter discusses established and ongoing developments in acquiring IR spectra of biomolecular ions, based on different spectroscopic approaches, available IR light sources, and molecular systems that are investigated.

4.2 Photodissociation with CO_2 Lasers

4.2.1 Background

CO_2 gas discharge lasers were the first IR lasers to be employed in conjunction with mass spectrometers to carry out IR multiple-photon dissociation (IRMPD) of trapped ions, as early as the late 1970s. CO_2 lasers are also currently the only routinely implemented lasers with commercial mass spectrometers. The turnkey operation and high continuous powers of CO_2 lasers lend themselves ideally to photodissociation experiments. As discussed in Chap. 2, CO_2 lasers are often operated at a fixed wavelength of 10.6 μm ($\approx$940 cm^{-1}), but the output wavelength can be tuned to other ro-vibrational transitions by virtue of a line-tunable diffraction grating.

4.2.2 Peptides

Much biomolecular MS in the past two decades has focused on peptides and proteins, due to the emergence of soft ionization methods, such as electrospray ionization (ESI) [1] and matrix-assisted laser desorption/ionization (MALDI) [2]. With these techniques, biological macromolecules, such as proteins, multiprotein complexes [3], and even entire viruses [4], can be transferred into the gas phase as intact entities. Measuring the masses of these species, however, only yields limited structural information. A key aspect in identifying peptides and proteins lies in confirming their sequences (i.e., primary structures). In a mass spectrometer, the peptide sequence is derived by dissociating the peptide into smaller fragments. Ion activation is conventionally achieved by accelerating the ion by electric fields and colliding it with a collision gas. This leads to vibrationally excited ions, which then undergo dissociation and hence incur a change in their mass-to-charge ratio (m/z). A similar heating of the ion can be achieved by absorbing IR photons. This begs the question how the IR spectrum of a peptide compares with the available output wavelengths of a CO_2 laser.

Figure 4.2 depicts IR photodissociation spectra of two naturally occurring peptides, bradykinin and Leu-enkephalin, ionized by potassium cation (K^+) adduction. These IR multiple-photon dissociation (IRMPD) "action" spectra were recorded by irradiating the trapped peptide ions with a tunable free-electron laser (see Chap. 2). The charge carrier K^+ is detached upon resonant absorption of multiple IR photons. The IR spectrum is obtained by monitoring the photodissociation yield as a function of wavelength. Diagnostic band positions are highlighted in Fig. 4.2, such as the backbone C=O stretching (amide I) and N–H bending (amide II) modes. The congested region from 1,000 to 1,450 cm^{-1} is composed of side-chain deformation modes, CH bending, and other less diagnostic modes. It is noteworthy that the fixed-frequency position of a CO_2 laser at 10.6 μm is relatively far removed from vibrational resonances. In fact, the entire accessible wavelength region of a line-tunable CO_2 laser shows little overlap with the inherent absorption spectra of these peptides.

Despite the reduced spectral overlap, irradiation of peptides with CO_2 lasers often does lead to abundant photodissociation. This is a result of the high continuous output power of CO_2 lasers, which can provide a hundred times more photon fluence than, for instance, free-electron lasers. In other words, even though 10.6 μm photons are merely absorbed in the tail of IR absorption bands, sufficiently high fluence can result in substantial photon absorption, leading to

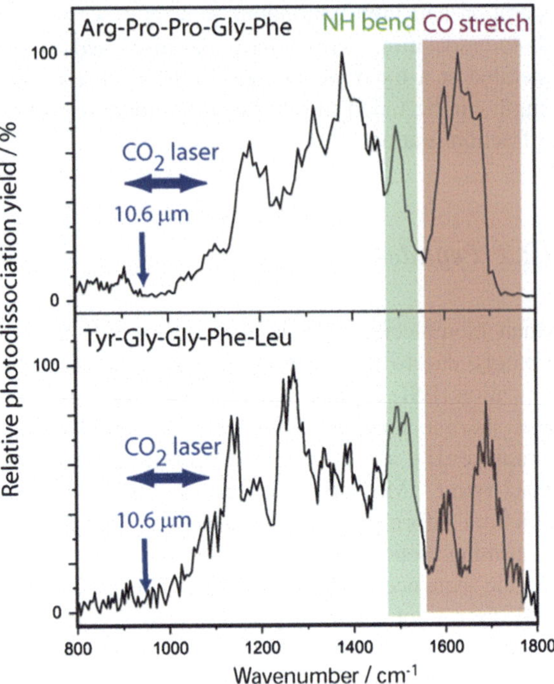

Fig. 4.2 Infrared spectra of peptides bradykinin fragment 1–5 (*top*) and Leu-enkephalin (*bottom*) ionized by K^+ attachment. Adapted with permission from Polfer et al. [27]. Copyright (2005) American Chemical Society

photofragmentation. On the other hand, this also explains why many peptides do not photodissociate at lower CO_2 laser power settings. Under lower-power conditions, only those molecules dissociate that resonantly absorb at the excitation wavelength.

In fact, some chemical moieties display considerable absorption at CO_2 laser output wavelengths. Phosphate groups exhibit intense P–O stretching and POH wagging in the 900–1,100 cm^{-1} region. The same applies to C–O stretch vibrations of saccharide (i.e., carbohydrate, glycan) moieties. Modified peptides, such as phosphorylated and glycosylated peptides, are hence excellent target molecules that are expected to display selective photodissociation near 10.6 μm. Figure 4.3 encapsulates the basic premise of the photodissociation approach for phosphorylated and glycosylated peptides, based on previous experimental results [5–7]. Photon irradiation at 10.6 μm results in selective photodissociation of phosphorylated and glycosylated peptides. The photofragmentation mass channels can give further insights into the biomolecular structures. For instance, phosphorylated peptides often display characteristic phosphate neutral losses (80 and 98 Da). Saccharide side-chain structures are often also more labile for cleavage, allowing a determination of the saccharide building blocks and their linkage patterns in branched saccharide chains.

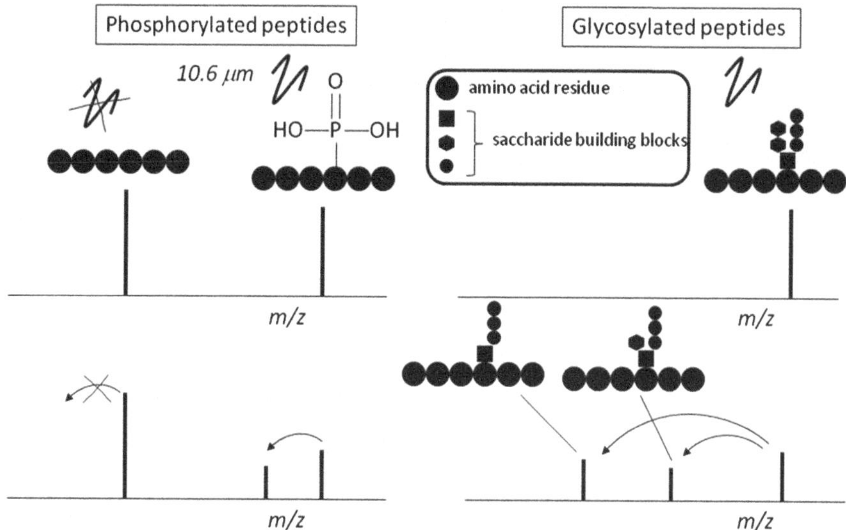

Fig. 4.3 Workflow showing selective photodissociation of phosphorylated and glycosylated peptides at 10.6 μm irradiation

In summary, the absence of vibrational resonances of most peptides in the tuning range of CO_2 lasers presents opportunities in selectively identifying those peptides that do include resonant absorbing moieties, such as phosphate or saccharide groups.

4.2.3 Saccharides

Elucidating the detailed structures of saccharides by MS presents a formidable analytical challenge, due to the large number of isomeric variants that need to be distinguished. Saccharides differ in the building blocks that make up these polymeric structures, as well as the linkage position and anomericity (i.e., chirality) of the glycosidic bond. Fixed-wavelength irradiation with CO_2 lasers has been employed for a number of years to derive information on saccharide structures [8], given their good absorbance characteristics near 10.6 μm.

Wavelength-variable IR photodissociation has not yet found the same level of application, despite promising results in distinguishing saccharide isomers [9, 10]. Figure 4.4 shows the IR spectra of a series of lithium cation (Li^+)-adducted disaccharides. These saccharides are isomers that differ in the linkage position and anomericity of the glycosidic bonds that link both glucose (Glc) building blocks. The structure of the β 1-6 linkage isomer is depicted. Note that the reducing end of the disaccharide is indicated as a wavy line, due to the mixture of α- and β-anomers in solution. The relative orientation of all other hydroxyl groups is fixed.

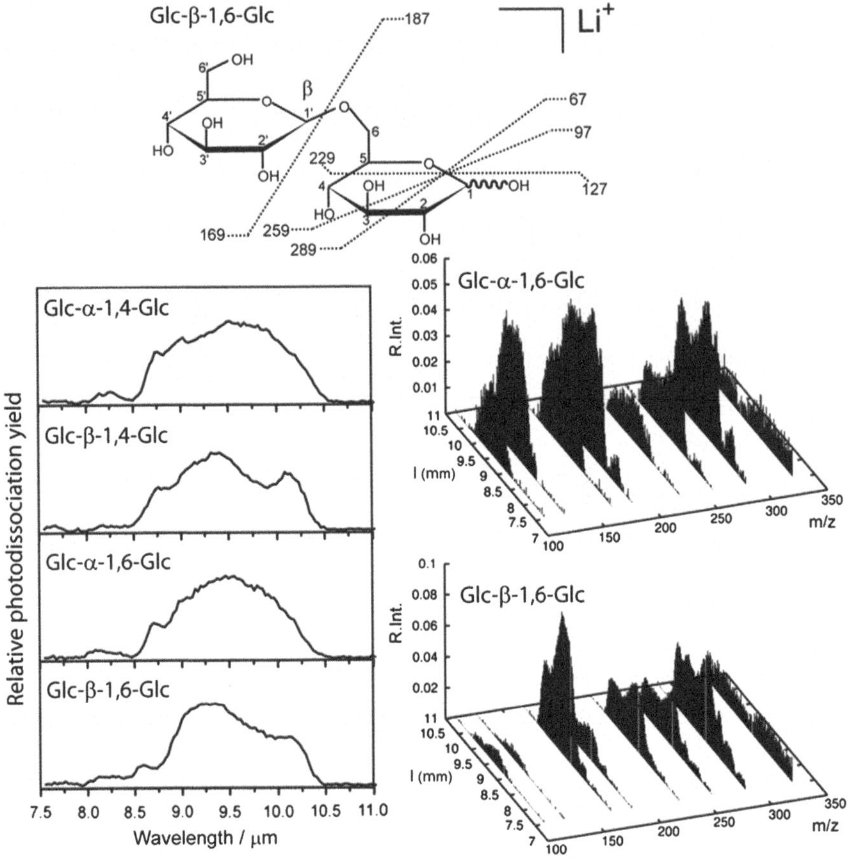

Fig. 4.4 *Top* Structure of Glc–Glc disaccharide, linked by β-1-6 glycosidic bond. Photodissociation mass channels are indicated. *Left* Infrared spectra of Li^+-adducted disaccharides. *Right* Individual photofragment mass channels as a function of laser wavelength. Adapted with permission from Polfer et al. [9]. Copyright (2006) American Chemical Society

The IR spectra in Fig. 4.4 display minimal differences, at first sight suggesting that this approach cannot differentiate between different isomers. However, instead of considering the sum of all photofragments as a function of wavelength, the wavelength dependence of each photodissociation mass channel can be investigated separately. In fact, these saccharides photodissociate into a multitude of fragment mass channels, as shown for the example of the Glc-β-1,6-Glc disaccharide. When plotting each photofragment mass channel separately, much more fine structure can be resolved in these photodissociation patterns. A comparison of these wavelength-dependent plots for different isomeric variants, such as for the α- and the β-anomers of the 1-6 linkage disaccharides, exhibit striking differences in their photofragmentation patterns. In fact, these photofragmentation patterns can be considered spectral fingerprints to allow identification.

While the data in Fig. 4.4 were acquired with a free-electron laser, more recently a benchtop, line-tunable CO_2 laser has been applied to the same task. Figure 4.5 shows a series of wavelength-dependent photodissociation patterns for different disaccharide isomers. Clearly, all linkage isomers and anomeric variants have characteristic photofragmentation patterns. Given the large information content in these multidimensional datasets, more sophisticated data analysis approaches are necessary to highlight the most significant differences [11]. Distinguishing between different isomers requires irradiation at some selected wavelengths, but not necessarily the whole range of CO_2 laser frequencies. The measurements of photodissociation at some selected wavelengths would greatly speed up the timescale of these measurements and hence reduce analysis time and sample consumption.

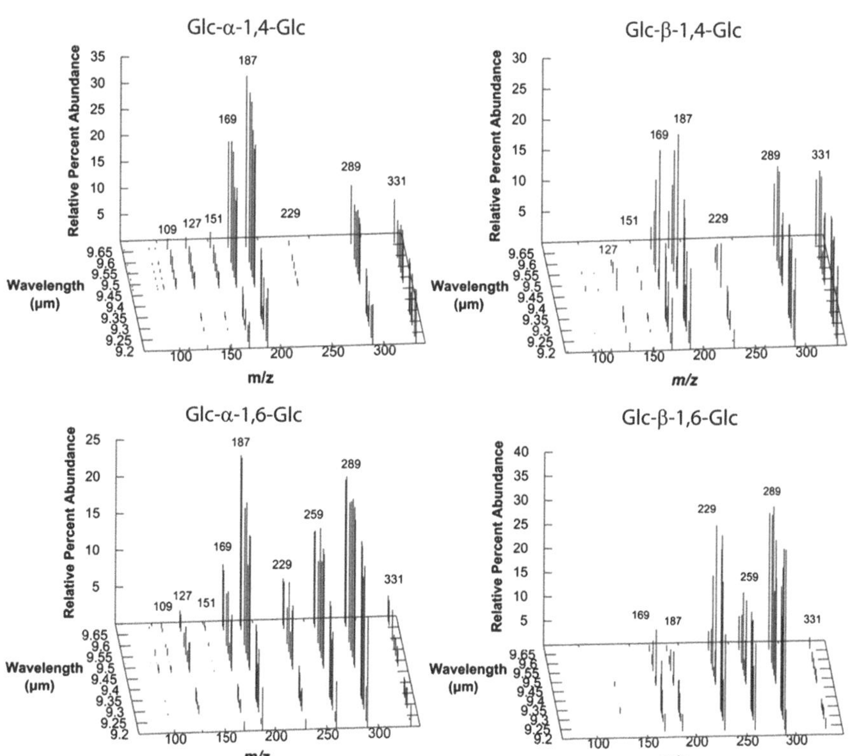

Fig. 4.5 Wavelength-dependent photodissociation patterns for 1-4 and 1-6 glucose–glucose disaccharides. Adapted with permission from Stefan et al. [10]. Copyright (2010) with permission from Elsevier

4.3 Photodissociation with Optical Parametric Oscillators

4.3.1 Background

Optical parametric oscillators/amplifiers (OPO/As) produce tunable idler radiation from 2,400 to 4,000 cm^{-1}, thus giving suitable coverage of the hydrogen-stretching region. The continuous output power of OPOs is significantly lower than that for CO_2 lasers, by about two orders of magnitude. On resonant absorptions, the output power of OPOs is typically sufficient to allow absorption of multiple photons. OPO irradiation can also be complemented with non-resonant irradiation from a CO_2 laser to enhance the photodissociation yield [12, 13]. This rests on the premise that resonant absorption of an OPO photon increases the density of states of the ion, increasing the absorption cross-section for the non-resonant absorption from the 10.6 μm radiation produced by the CO_2 laser [14, 15].

4.3.2 Peptides

A number of IR-active moieties that are present in peptides light up in the hydrogen-stretching region. The amide NH, indole NH, carboxylic acid OH, alcohol OH, and phosphate OH stretches can typically be distinguished based on diagnostic band positions. Figure 4.6 summarizes experimental IR spectra of various amino acids and peptides, where the diagnostic bands are highlighted by color-coding. These band positions are also summarized in Table 4.1. A particularly useful application is the identification of phosphorylated peptides, based on the phosphate OH stretch from 3,650 to 3,700 cm^{-1} [16]. In this way, phosphorylated peptides can be conveniently screened in peptide mixtures, as only these peptides will photodissociate at the resonant phosphate OH stretching frequency [17].

Peptides are most commonly ionized by proton attachment on basic sites, such as the side chains of arginine or lysine residues, or on the N-terminal amine. Protonated amines (i.e., NH_3^+) have a strong binding affinity for crown ethers and 18-crown-6 ether (18c6) in particular. The 3-fold axis of symmetry of NH_3^+ allows three strong hydrogen bonds with oxygen atoms on the 6-fold order of symmetry of 18c6. Non-covalent complexes of 18c6-adducted protonated peptides can be readily generated in the electrospray process [18]. Laser irradiation of these complexes often results in exclusive loss of the crown ether. The complexation approach thus has a number of advantages for IR photodissociation spectroscopy:

- The efficiency and sensitivity of IR photodissociation is enhanced, as the photodissociation yield is concentrated on one fragment mass channel, as opposed to being diluted over multiple channels.

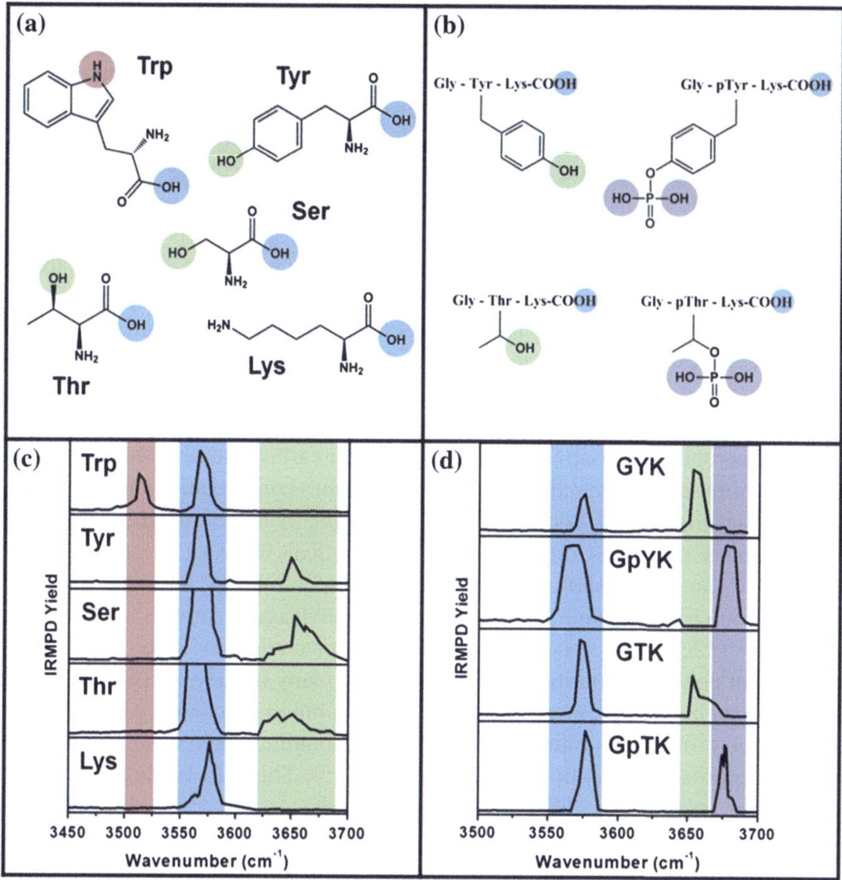

Fig. 4.6 Infrared spectra in the hydrogen-stretching region, showing diagnostic vibrations. Adapted with permission from Stedwell et al. [28] and [17]. Copyright (2012 and 2013) American Chemical Society

Table 4.1 Summary of diagnostic hydrogen-stretching band ranges from Fig. 4.6

Hydrogen-stretching modes	Range/cm^{-1}
Phosphate OH	3,665–3,685
Phenol/alcohol OH	3,640–3,670
Carboxylic acid OH	3,550–3,590
Indole NH	3,500–3,530

- The IR spectra of multiple precursor ions can be recorded simultaneously in a multiplexed approach, as each complex dissociates into a separate and unique mass channel.

The basic premise of the multiplexing approach to effecting IR photodissociation on multiple analytes in parallel is illustrated in Fig. 4.7. Multiple 18c6-complexed peptides are irradiated by IR photons at discrete wavelengths. The photon-induced loss of the crown ether proceeds into separate mass channels for each peptide complex, thus permitting unambiguous identification of chemical moieties in various peptides. For this fictitious example, peptides with phosphate OH, alcohol OH, and indole NH moieties can be distinguished based on diagnostic frequencies. At non-resonant frequencies, none of the peptides photodissociate.

A potential caveat in this approach is the effect of the binding partner on the IR spectra. This effect is demonstrated for the case of protonated tryptophan, $TrpH^+$, in Fig. 4.8. The NH stretching modes associated with the NH_3^+ moiety (yellow) are significantly redshifted to the 3,100–3,200 cm^{-1} region upon complexation with 18c6. This, in addition with the intense CH stretching modes (dark gray), makes the 2,800–3,250 cm^{-1} range highly congested and minimizes the diagnostic value of this region. In contrast, the local oscillator carboxylic acid OH and phenol OH stretching modes remain well resolved upon complexation and are only weakly blueshifted (5–10 cm^{-1}). This predictable behavior of the higher-frequency modes maintains the considerable analytical value of these vibrations, while at the same time allowing multiplexed IR spectroscopy measurements.

Other non-covalent binding partners can be envisaged. Another crown ether, 12-crown-4 (12c4), undergoes similar photofragmentation to 18c6, but displays less efficient complexation in the ESI process for many peptides. Even 18c6 is far from the ideal binding partner for multiplexed IR photodissociation, as it is limited to binding to protonated amines. More ubiquitous binding partners are required for a general approach to multiplexed IR spectroscopy. This will be discussed again below in the section on cold spectroscopy methods.

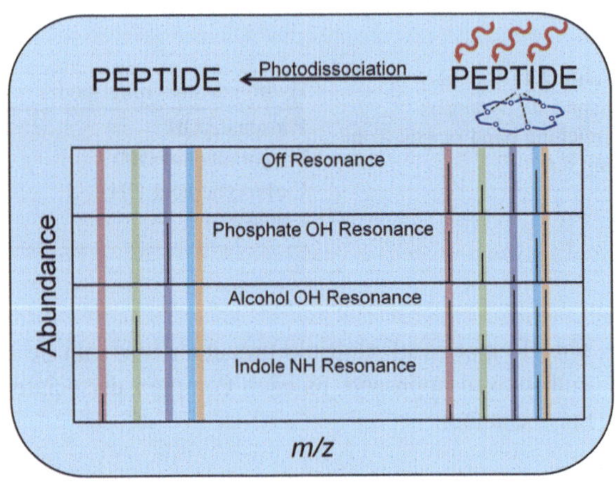

Fig. 4.7 Multiplexing of infrared photodissociation experiment, where the presence of various chemical moieties can be verified based on diagnostic band positions. Adapted with permission from Stedwell et al. [17]. Copyright (2012) American Chemical Society

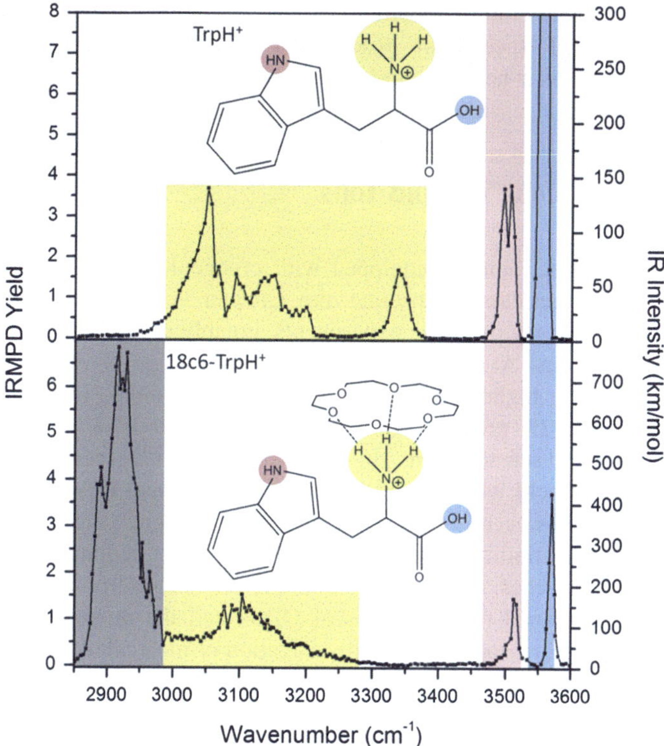

Fig. 4.8 Infrared spectra of bare protonated tryptophan (*top*) and 18c6-adducted protonated tryptophan (*bottom*). The IR bands are color coded: carboxylic acid OH (*blue*), indole NH (*red*), NH_3^+ modes (*yellow*), and 18c6 CH stretches (*dark gray*). Adapted with permission from Stedwell et al. [28]. Copyright (2013) American Chemical Society

4.3.3 Saccharides

Saccharides are rich in alcohol OH moieties. While the C–O stretch modes in the mid-IR range give rise to broad, unresolved features (see Fig. 4.3), the local oscillator nature of OH stretching modes may hold more promise in distinguishing isomeric variants. The IR spectra of two saccharide epimers, glucuronic acid and iduronic acid complexed with rubidium cation (Rb^+), are compared in Fig. 4.9. These isomers differ in the chirality of carbon 5, thus affecting the orientation of the carboxylic acid group. The favored computed geometries for both cation-bound isomers are also depicted in Fig. 4.9. The binding motifs are clearly different, thus rationalizing the differences in IR spectra. In particular, the structural differences between both isomers are found to play a crucial role in how effectively the Lewis base oxygen atoms on the saccharide ring can chelate the metal cation. This in turn affects the hydrogen-bonding interactions of the OH groups with each other and consequently the IR spectra of these complexes. It should be noted that the differences in IR spectra are exacerbated by the metal complexation. The choice of metal cation is hence an important factor in discriminating these isomers.

4.4 IR spectroscopy of Cold Ions

In custom mass spectrometers equipped with cryogenic ion traps, ions can be cooled to temperatures close to absolute zero. Chapter 3 elaborated on some of the ion trap devices to carry out these experiments, notably cryogenic quadrupole ion traps and 22-pole traps. At sufficiently cold temperatures, gas-phase conformations are frozen, allowing higher-resolution spectroscopic methods. Examples of two techniques, namely IR predissociation spectroscopy and IR–UV ion dip spectroscopy, will be discussed here. In contrast to IRMPD spectroscopy, these methods typically yield well-defined, narrow IR bands, and they are compatible with the reduced-power mid-IR (800–2,100 cm^{-1}) output ranges of $AgGaSe_2$ crystals. This compatibility arises from the lower-power requirements of these single-photon approaches, as opposed to multiple-photon absorption in IRMPD. In summary, the combined output ranges of OPOs and the downconversion with $AgGaSe_2$ crystals allow spectroscopic investigation of molecules in the hydrogen-stretching and mid-IR regions using benchtop tunable light sources.

Figure 4.10 depicts results using the IR predissociation method for the example of a small peptide. The cryogenically cooled peptide ions are tagged with a weakly bound atom or molecule, in this case molecular hydrogen (H_2). In fact, the mass spectra confirm that under these conditions, up to four H_2 molecules were attached.

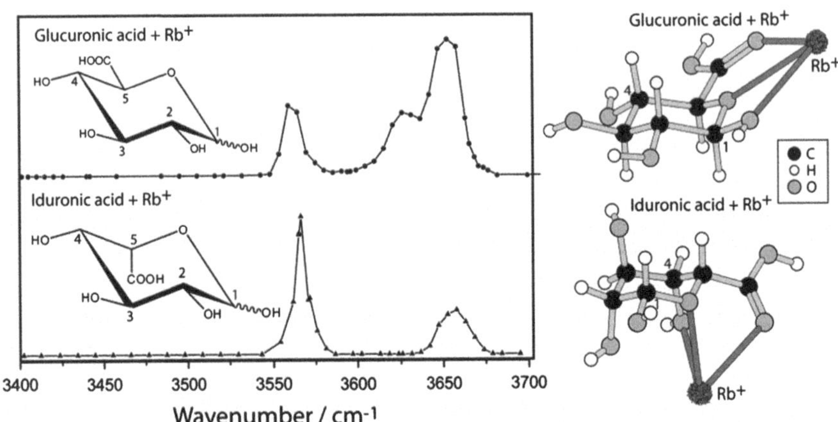

Fig. 4.9 *Left* Infrared spectra of saccharide isomers glucuronic and iduronic acids complexed with Rb^+. *Right* Best computed configurations, showing complexation with Rb^+. Adapted from Cagmat et al. [29]

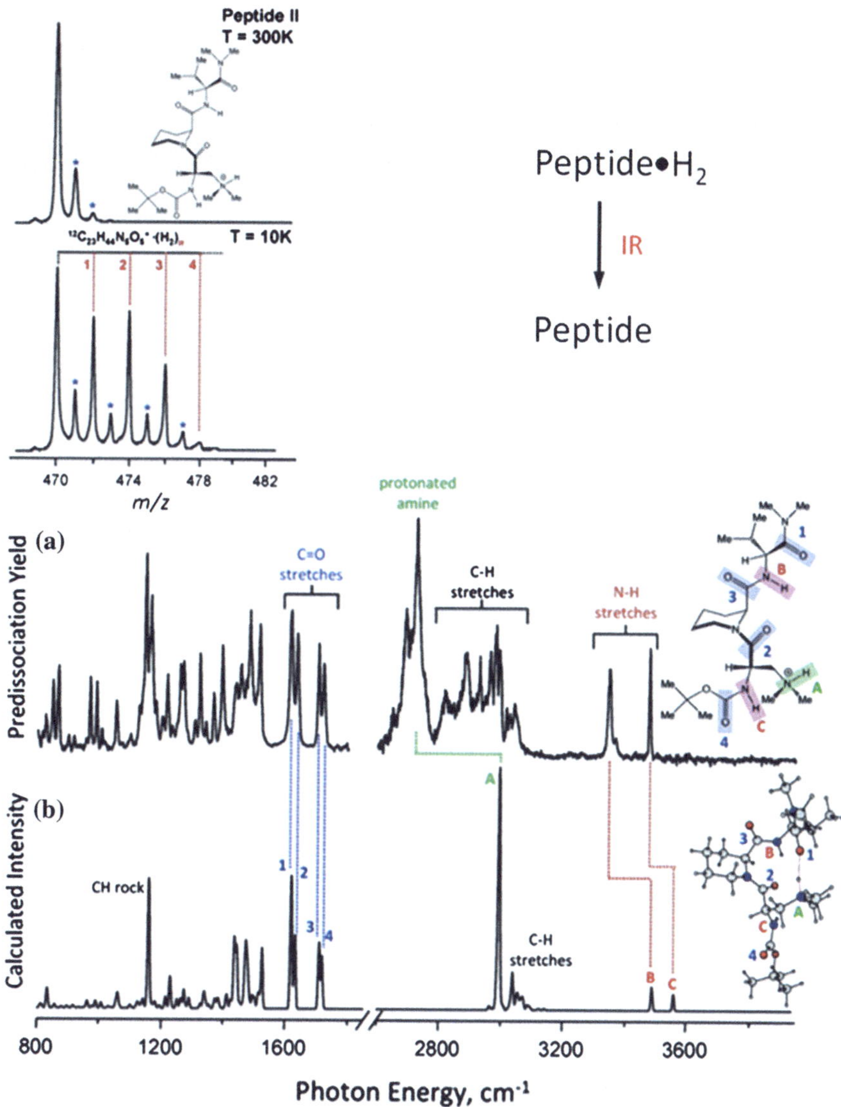

Fig. 4.10 *Top left* Tagging of peptide with H_2 molecules at cryogenic temperatures (10 K). *Top right* Schematic of IR predissociation of tagged peptide via loss of H_2 tag. *Bottom* Infrared spectrum of cryogenically cooled peptide using IR predissociation technique compared to computed IR spectrum. Adapted with permission from Kamrath et al. [30]. Copyright (2011) American Chemical Society

In IR predissociation spectroscopy, the weakly bound tag is detached upon resonant IR absorption. Due to the low binding energy of the van der Waals-bound molecule, absorption of a single photon is sufficient to facilitate loss of the tag.

The IR predissociation spectrum of the peptide shown in Fig. 4.10 was recorded by mass isolating peptide •H_2 and monitoring the loss of the H_2 tag as a function of wavelength. As the IR spectrum of the tagged peptide is recorded, as opposed to the bare peptide, this begs the question whether the presence of the tag affects or even distorts the IR spectrum. The H–H stretching mode of H_2 is typically observed at >4,000 cm^{-1}, limiting potential overlap with other diagnostic bands. Some tags (e.g., Ar) have been shown to affect high-amplitude modes in proton-bound molecular systems [19]. This is, however, rather an exception to the rule. A number of studies have shown that the tag barely affects the IR spectra. In the example in Fig. 4.10, the peptide bands are well resolved, showing for instance a clear distinction between the four backbone C=O stretching modes in this molecule in the 1,600–1,700 cm^{-1} range. A comparison of this experimental spectrum with a calculated spectrum by quantum-chemical approaches shows that some spectral features are well reproduced, whereas others are not. The agreement between experiment and theory is certainly far from perfect, notably in the hydrogen-stretching region, even if a qualitative assessment of the conformation is confirmed. In order to validate band assignments, an isotope labeling strategy is often employed. This approach is illustrated in Fig. 4.11 for the example of an amide I band. Selective carbon isotope substitutions from ^{12}C to ^{13}C result in a characteristic redshift of the C=O stretch. This strategy works particularly well in vibrational modes that can be considered local oscillators, as the change in reduced mass is mainly constrained to this vibration.

Similar high-resolution IR spectra can be obtained using the IR–UV dip spectroscopy technique [20]. Figure 4.12 depicts results for a heptapeptide that has a high propensity to form alpha-helical structures. The method rests on the premise

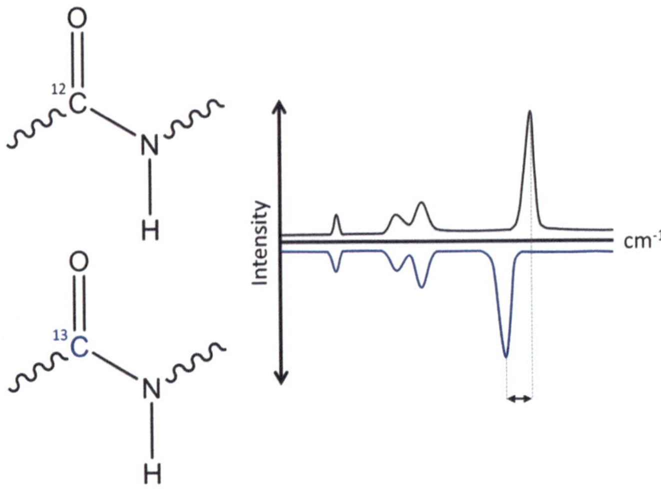

Fig. 4.11 Effect of isotope shift for the example of peptide amide *I* band, upon going from ^{12}C=O to ^{13}C=O. The spectrum for ^{13}C=O is pointing downward to facilitate comparison

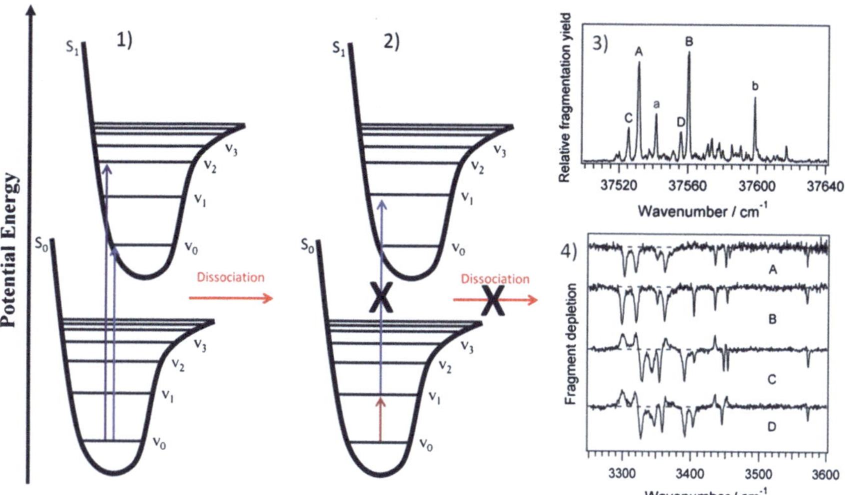

Fig. 4.12 UV photodissociation spectrum of cryogenically cooled Ac-Phe-(Ala)$_5$-Lys-H$^+$. By parking the UV laser on a particular electronic resonance, the infrared spectra of selected conformers (A, B, C, or D) can be recorded in turn using the IR–UV ion dip technique. Adapted from Stearns et al. [31] with permission from the PCCP Owner Societies

that the molecule interacts with both IR and UV photons, and hence, vibrational and electronic spectroscopy measurements are taken. Absorption of a UV photon requires the presence of a UV chromophore to undergo an $S_0 \rightarrow S_1$ transition, in this case the phenylalanine aromatic side chain, which exhibits a $\pi \rightarrow \pi^*$ transition. Absorption of a UV photon leads to photodissociation, either by coupling to an electronic repulsive state or via internal conversion to highly vibrationally excited ions on the ground electronic state (see Chap. 1). Crucially, the $\pi \rightarrow \pi^*$ transition is dependent on the orientation of the phenylalanine side chain and hence the conformation of the molecule. In the UV photodissociation spectrum in Fig. 4.12, a number of peaks are observed. These peaks correspond to different $S_0 \rightarrow S_1$ transitions, either due to the presence of different conformations or due to different vibrational–electronic transitions for the same conformers. For the cold ions in these experiments, only the ground vibrational states are expected to be populated (i.e., $\mathbf{v}_0$ for every normal mode). In the UV photon absorption process, excited vibrational states (i.e., $\mathbf{v}_1$, $\mathbf{v}_2$) of S_1 can be accessed. These so-called vibronic bands occur at higher frequencies, due to the larger energy gap for instance for $\mathbf{v}_0 \rightarrow \mathbf{v}_2$, as opposed to the origin band $\mathbf{v}_0 \rightarrow \mathbf{v}_0$. In Fig. 4.12, capitals A–D designate the origin bands for different conformers, whereas the lowercase letters a and b refer to vibronic bands for conformers A and B, respectively.

Despite the power of UV spectroscopy in separating different conformations, detailed structural information on these conformers is often difficult to ascertain by these measurements. Conversely, an IR spectrum of a particular conformer can serve as a strong structural identifier. By setting the UV laser on a particular transition, for instance at 37,532 cm^{-1}, structure A is selected. The IR spectrum of structure A is then obtained by scanning the IR frequencies. In fact, the IR pulse precedes the UV laser pulse. Resonant IR absorption leads to a depletion of $\mathbf{v}_0$ of the electronic ground state. This population transfer thus accounts for a decrease in the origin band for conformer A. This is why IR bands in Fig. 4.12 point downward, as seen for conformers A and B. On the other hand, in the IR spectra for conformers C and D, some of the bands point upward. This represents an increase in UV photodissociation yield upon IR photon absorption. These increases are observed at frequencies that match the IR spectra for A and B, and it is hence thought that these gains result from conformers A and B. The appearance of conformers A and B at a selective UV frequency for conformer C shows that the UV selection scheme is not always discriminatory, at least in cases when preceded by IR photon absorption. Nonetheless, the upward versus downward bands in IR–UV ion dip spectroscopy allow a simple distinction between conformer-selective and non-selective bands at a particular UV transition.

A comparison of the experimental data with computed IR spectra using quantum-chemical approaches (e.g., density functional theory) can yield detailed structural insights. The structural information from these measurements is summarized in Fig. 4.13. While all of the observed conformations display types of alpha-helical structures, there are some subtle differences in their hydrogen-bonding interactions. These differences can be ascertained by the positions of the NH stretches, as the strengths of hydrogen-bonding interactions are intimately linked to the extent of redshifting of these bands. An individual assignment of NH stretches can once again be achieved by isotopic labeling, in this case ^{14}N to ^{15}N, leading to characteristic redshifts of $\sim$8 cm^{-1}, as expected for a local oscillator mode. Each conformation thus provides a distinctive signature of NH stretching modes. An important caveat in UV spectroscopy is not to directly equate UV intensities with relative abundances of conformers. The relative UV intensities are of course determined by their respective abundances, but, crucially, are also dependent on the respective Franck–Condon (F–C) factors of these transitions. It is thus possible that a relatively sparse conformer exhibits an extremely efficient transition and hence a strong UV resonance, but that a much more abundant conformer is suppressed due to a poor F–C overlap.

IR–UV ion dip spectroscopy is a tremendously powerful tool in identifying different gas-phase conformations, due to its ability to provide IR spectra of selected conformations. It should be noted that the selection of conformers is also possible with an IR-only spectroscopic scheme. In a variation on hole-burning spectroscopy, in IR–IR double-resonance spectroscopy [21], a fixed-frequency IR photon serves to deplete away a particular conformation at a resonant frequency; a second wavelength-tunable IR photon then probes the IR spectrum of the conformers that remain. In other words, instead of selecting one conformation, as performed in IR–UV ion dip spectroscopy, one conformer (or more conformers) is (are) resonantly "burned" away.

IR spectroscopic measurements of cold ions stand at the forefront of structural information that is obtainable on mass-selected ions. These and similar techniques

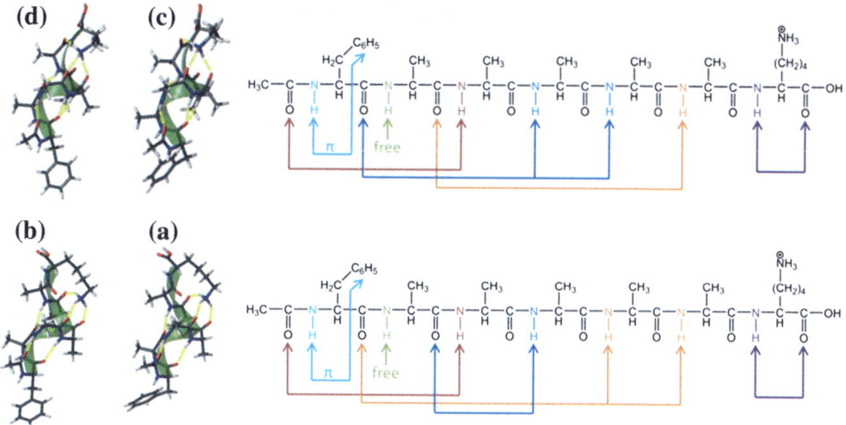

Fig. 4.13 Hydrogen-bonding network and geometries for conformers (**d**, **c**, **b**) and (**a**). Adapted from Stearns et al. [31] with permission from the PCCP Owner Societies

have already been demonstrated on peptides, peptide–ligand systems [22], peptide fragmentation products [23], saccharides and can be extended to many other molecular systems. The high resolution and conformational information from these measurements allow many interesting biophysical problems to be addressed, such as the inherent secondary structures of peptide motifs, as well as the effect of metal chelation or solvation/hydration on their structures [24]. In addition, the interactions between non-covalently bound ligands can be investigated with these approaches [22].

As detailed structural assignments require comparison with computed IR spectra from quantum-chemical approaches, another main benefit from these experiments is the rigorous testing of theoretical methods. The lack of agreement between experiment and theory is often caused by limitations in the theory and the harmonic oscillator approximation in particular. The availability of conformationally resolved IR spectra thus plays an important role in benchmarking calculated IR spectra.

4.5 Longer-Term Considerations for IR Spectroscopy of Mass-separated Biomolecules

4.5.1 Current Constraints

Many of the experimental approaches described in this chapter are currently limited to a few laboratories around the world. Nonetheless, as the increased structural information from IR spectroscopy measurements becomes more widely publicized, there will be a demand for extending these techniques to the wider scientific community.

Biophysical applications of IR spectroscopy are currently ongoing, and examples have already been highlighted in the section on cold spectroscopy methods. Conversely, bioanalytical applications of IR spectroscopy are developed to a much lesser extent, despite a huge potential in enhancing the structural information from MS measurements. In order to develop the technique as a routine tool in MS-based analysis of unknowns, as illustrated in Fig. 4.1, a number of current constraints need to be overcome are

- The operation of tunable IR light sources remains challenging. The ongoing development of tunable benchtop light sources that can truly be operated as turnkey systems will facilitate this process.
- Commercially available MS instrumentation is not designed or optimized for conducting laser spectroscopy experiments. The commercial availability of instrumentation that includes for instance cryogenic ion traps will be essential to maximize the structural information from these approaches.

Both high- and low-resolution spectroscopic techniques are likely to play a role in extending the structural information for gas-phase biomolecular ions, in analogy with high- and low-resolution techniques in MS.

4.5.2 Future Directions

IR spectroscopy is a powerful structural technique for smaller and moderately sized molecules. For large molecules, such as proteins, the experimental approaches become much more challenging, and the IR spectra become highly congested, thus complicating the structural interpretation [25].

Given the limited studies thus far, it is perhaps useful to speculate on the potential applications of IR spectroscopy of ions. Here, we attempt to demonstrate the potential usefulness of IR spectroscopy as a bioanalytical tool for the example of metabolites. Metabolites are involved in the metabolism of living creatures and thus affect a vast range of cellular activities, ranging from cell growth, signaling, to reproduction. Chemically, metabolites are categorized as small molecules, but they also encompass many classes of molecules, including amino acids, peptides, lipids, and carbohydrates. This is in principle an ideal combination for IR spectroscopy. On the one hand, as relatively small molecules, metabolites are expected to exhibit limited spectral congestion, thus maximizing the specific structural information that can be obtained. On the other hand, many chemical classes of metabolites are expected to have considerable differences in their chemical functionalities and are thus amenable to be distinguished by IR spectroscopy.

We have considered the >1 million metabolites from the chEMBL database [26] to derive statistical trends. An example of five isomers at exact mass

400.1205 amu (elemental composition: $C_{19}H_{20}N_4O_4S$) is depicted in Fig. 4.14. While the mass information cannot distinguish these molecules, the IR information is expected to be markedly different. Merely considering the alcohol OH and ether moieties, there are considerable differences between most of these variants. Many additional differences are apparent, such as the S=O bonds for isomer 1.

The subset of isomers above serves as a useful example for the potential application of IR spectroscopy in telling isomers apart. The capabilities to distinguish many metabolites by virtue of mass and IR information are illustrated in Fig. 4.15. The entries in the chEMBL database are plotted as a function of mass and IR information. The IR information is given in a binary format, where various functional modes are either present (1) or absent (0). A similar distinction had already been shown in Fig. 4.1. Only the ten most abundant IR bins are represented here, encompassing under 50 % of the entries; the remaining entries are distributed over 620 bins (not shown). The additional IR dimension in Fig. 4.15 illustrates the power of IR spectroscopy in separating a vast number of these analytes. In fact, the binary format does not even reflect the true magnitude in distinguishing these structures, as it ignores the number of times that a particular moiety is present in a molecule. As shown for the peptide in Fig. 4.10, the four individual backbone C=O stretches could be distinguished. However, resolving

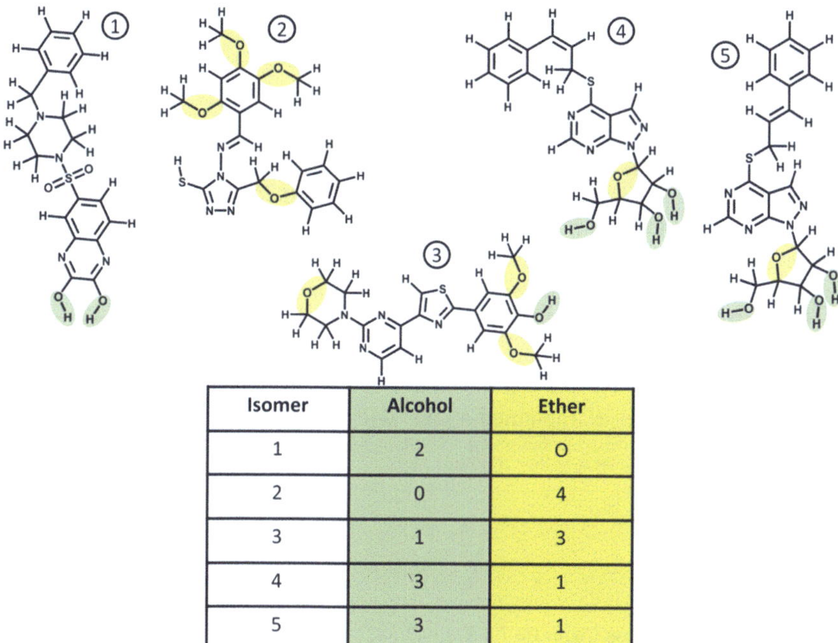

Isomer	Alcohol	Ether
1	2	0
2	0	4
3	1	3
4	3	1
5	3	1

Fig. 4.14 Isomeric metabolite molecules 1–5 at mass 400.1205 amu, showing color-coded IR-active alcohol (*green*) and ether (*yellow*) moieties. The number of these moieties per molecule is tabulated

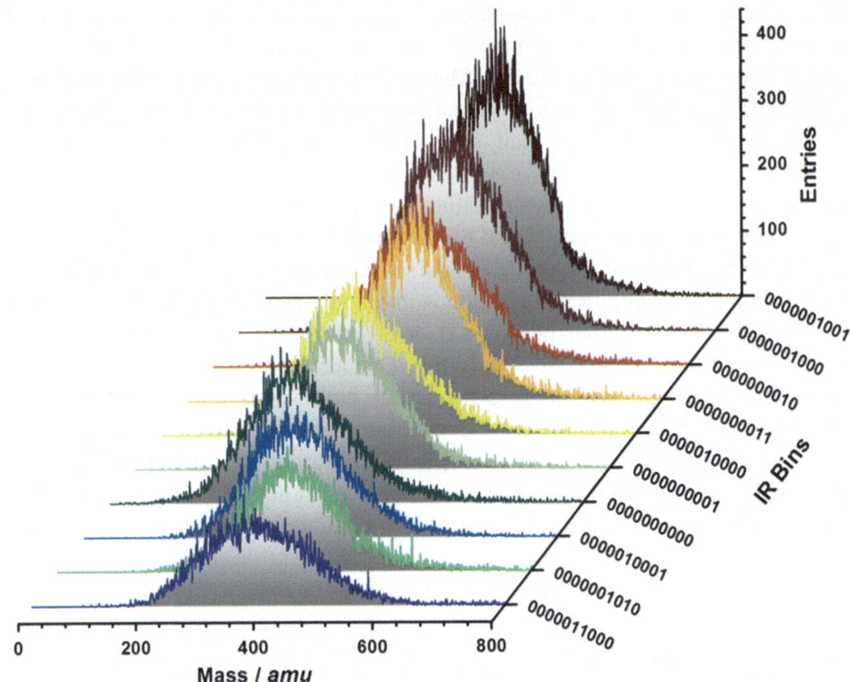

Fig. 4.15 Separation of metabolite entries in chEMBL database as a function of nominal mass and IR information. Adapted from Stedwell et al. [32]

these more subtle features requires higher-resolution spectroscopy approaches and thus cryogenic MS instrumentation to probe cold ions.

Many more bioanalytical applications of IR spectroscopy can be foreseen. For instance, isomeric variants pose a considerable challenge in the analysis of saccharides. IR spectroscopy is also well placed for targeted studies, where only selected classes of molecules are identified. Phosphopeptides are a good case in point, based on selective photodissociation at the phosphate OH stretching frequency. Many of these implementations will likely become routine over the coming decade.

References

1. Fenn JB (2003) Angew Chem Int Ed Engl 42:3871–3894
2. Karas M, Bahr U, Giessmann U (1991) Mass Spectrom Rev 10:335–357
3. Ruotolo BT, Giles K, Campuzano I et al (2005) Science 310:1658–1661
4. Uetrecht C, Rose RJ, van Duijn E et al (2010) Chem Soc Rev 39:1633–1655
5. Flora JW, Muddiman DC (2001) Anal Chem 73:3305–3311

6. Crowe MC, Brodbelt JS (2005) Anal Chem 77:5726–5734
7. Hakansson K, Cooper HJ, Emmett MR et al (2001) Anal Chem 73:4530–4536
8. Tseng K, Hedrick JL, Lebrilla CB (1999) Anal Chem 71:3747–3754
9. Polfer NC, Valle JJ, Moore DT et al (2006) Anal Chem 78:670–679
10. Stefan S, Eyler JR (2010) Int J Mass Spectrom 297:96–101
11. Stefan S, Ehsan M, Pearson WL et al (2011) Anal Chem 83:8468–8476
12. Mino WK Jr, Szczepanski J, Pearson W et al (2010) Int J Mass Spectrom 297:131–138
13. Sinha RK, Erlekam U, Bythell B et al (2011) J Am Soc Mass Spectrom 22:1645–1650
14. Yeh LI, Price JM, Lee YT (1989) J Am Chem Soc 111:5597–5604
15. Peiris DM, Cheeseman MA, Ramanathan R et al (1993) J Phys Chem 97:7839–7843
16. Scuderi D, Bakker JM, Durand S et al (2011) Int J Mass Spectrom 308:338–347
17. Stedwell CN, Patrick AL, Gulyuz K et al (2012) Anal Chem 84:9907–9912
18. Julian RR, Beauchamp JL (2001) Int J Mass Spectrom 210(211):613–623
19. Hammer NI, Diken EG, Roscioli JR et al (2005) J Chem Phys 122:244301
20. Rizzo TR, Stearns JA, Boyarkin OV (2009) Int Rev Phys Chem 28:481–515
21. Leavitt CM, Wolk AB, Fournier JA et al (2012) J Phys Chem Lett 3:1099–1105
22. Garand E, Kamrath MZ, Jordan PA et al (2012) Science 335:694–698
23. Wassermann TN, Boyarkin OV, Paizs B et al (2012) J Am Soc Mass Spectrom 23:1029–1045
24. Kamariotis A, Boyarkin OV, Mercier SR et al (2006) J Am Chem Soc 128:905–916
25. Oomens J, Polfer N, Moore DT et al (2005) Phys Chem Chem Phys 7:1345–1348
26. Gaulton A, Bellis LJ, Bento AP et al (2012) Nucleic Acids Res 82:2456–2462
27. Polfer N, Paizs B, Snoek LC et al (2005) J Am Chem Soc 127:8571–8579
28. Stedwell CN, Galindo JF, Gulyuz K et al (2013) J Phys Chem A 117:1181–1188
29. Cagmat E, Szczepanski J, Pearson W et al (2010) Phys Chem Chem Phys 12:3474–3479
30. Kamrath MZ, Garand E, Jordan PA et al (2011) J Am Chem Soc 133:6440–6448
31. Stearns JA, Seaiby C, Boyarkin OV et al (2009) Phys Chem Chem Phys 11:125–132
32. Stedwell CN, Galindo JF, Roitberg A et al. (2013) Ann Rev Anal Chem 6:267–285

Chapter 5
UV-Visible Activation of Biomolecular Ions

Rodolphe Antoine and Philippe Dugourd

5.1 Electronic Excitation of Biomolecular Ions

Excited levels involved in photoexcitation of molecular ions depend on the wavelength of the light (see Fig. 5.1). Infrared photons correspond to vibrational excitation and lead to a global heating of ions and fragmentation on the ground state potential energy surface. In this chapter, we will discuss visible and UV excitations that occur in biomolecular ions. Visible and UV photons correspond to electronic excitations of valence electrons. At higher energy photons, core-shell electron excitation is possible. Figure 5.2 displays the absorption spectrum recorded for cytochrome C protein in water. Above 320 nm, the absorption is due to the electronic excitation of the prosthetic heme (porphyrin) group, in particular the band observed at 410 nm corresponds to the Soret Band of the heme molecule. Around 260 nm, the absorption is due to the electronic excitation of aromatic amino acids (tryptophan, tyrosine and phenylalanine). Below, 220 nm, the absorption from the protein backbone, in particular from peptide bonds, is observed.

In the case of linear excitation (absorption of a single photon), the energy deposited in the system is well-defined and corresponds exactly to the photon energy. After population of the excited electronic state, different de-excitation pathways are possible (see Fig. 5.3):

- Emission of a photon (fluorescence).
- Internal conversion (IC) from electronic energy to vibrational energy, followed by internal vibrational redistribution (IVR).
- Fragmentation in electronic excited state.
- Electron emission (depending on photon energy).

R. Antoine · P. Dugourd (✉)
Institut Lumière Matière, UMR5306 Université Lyon 1-CNRS, Université de Lyon, 69622 Villeurbanne cedex, France
e-mail: philippe.dugourd@univ-lyon1.fr

N. C. Polfer and P. Dugourd (eds.), *Laser Photodissociation and Spectroscopy of Mass-separated Biomolecular Ions*, Lecture Notes in Chemistry 83,
DOI: 10.1007/978-3-319-01252-0_5,

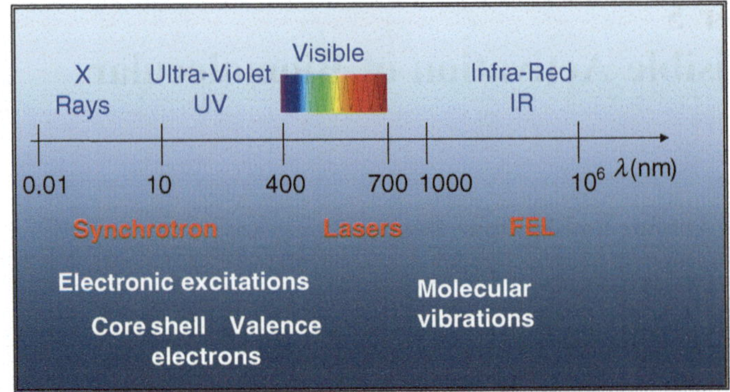

Fig. 5.1 Decomposition of the electromagnetic spectrum according to the wavelength from 0.01 to 10^6 nm. (*FEL* Free electron laser)

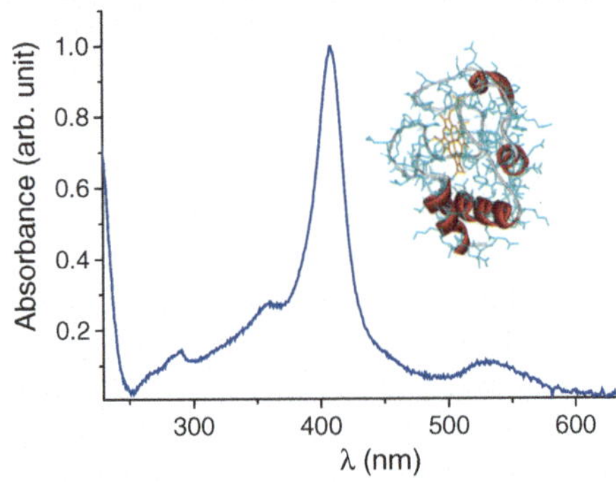

Fig. 5.2 Absorption spectrum of bovine Ferri-Cytochrome c in water. The pdb X-ray structure of the protein is given in *inset*. The heme molecule is highlighted in *orange*

The coupling between optical UV-visible spectroscopy with mass spectrometry allows us to monitor several de-excitation pathways. In particular, fragmentation occurring in electronic excited states and fragmentation in the ground state after IC and IVR.

5.2 Gas-Phase Visible and Near-UV Excitation of Biomolecular Ions

For anions, electron emission is often observed after irradiation with UV/visible photons [1]. Monitoring electron detachment yield permits to record action spectra of trapped ions. Figure 5.4 shows the action spectra recorded for two proteins. Figure 5.4b presents the UV spectrum of gas-phase α-lactalbumin. Electron detachment is observed below 300 nm. Detachment yield is directly proportional to the laser fluence which shows that electron detachment is a one-photon process.

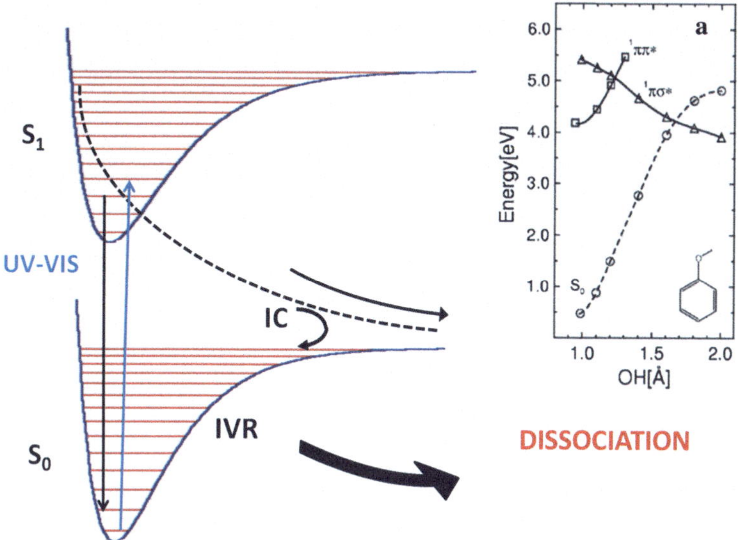

Fig. 5.3 Simplified diagram of photoexcitation and de-excitation pathways. The absorption of UV/VIS photon energy (*blue arrow*) leads to the conversion of electrons to an excited electronic state. Possible relaxation pathways are: photon emission (*black arrow*), coupling to a dissociative state or internal conversion (IC) with coupling to the ground state (*dotted line*). Statistical redistribution after internal conversion can be followed by unimolecular dissociation in the ground electronic state. *Right panel* Calculated potential energy (PE) profiles for phenol of the lowest $^{1}\pi$–π^{*} states (*squares* and *diamonds*), the lowest $^{1}\pi$–σ^{*} state (*triangles*) and the electronic ground state (*circles*) as a function of the OH stretch (phenol) reaction coordinate. *Via* predissociation of the $^{1}\pi$–π^{*} states and a conical intersection with the ground state, the $^{1}\pi$–σ^{*} states trigger an ultrafast internal-conversion process, which is essential for the photostability of biomolecules. Insert reproduced/adapted from Ref. [24] with permission of the PCCP Owner Societies

A band centered at 280 nm corresponds to π–π^{*} excitations of aromatic amino acids, while the increase observed below 240 nm is due to both chromophores and protein backbone [2]. The electronic spectroscopy of aromatic amino acids was pioneered by Levy and coworkers in 1985 [3].

While most of biomolecular ions do not absorb below 300 nm, visible absorption is observed in the presence of visible chromophore as for example prosthetic group in proteins. Figure 5.4a shows the visible region of the action spectrum of gas-phase cytochrome *c*. This spectrum should be compared to Fig. 5.2. The Soret band is observed at 410 nm both in gas-phase and solution spectra, showing that the native structure of cytochrome c can be preserved in the gas phase. Detachment yield is proportional to the laser fluence which shows that electron detachment is a one-photon process.

For biomolecules without natural visible chromophores, a strategy aiming at using derivatization with visible dyes is particularly suitable. This allows to use

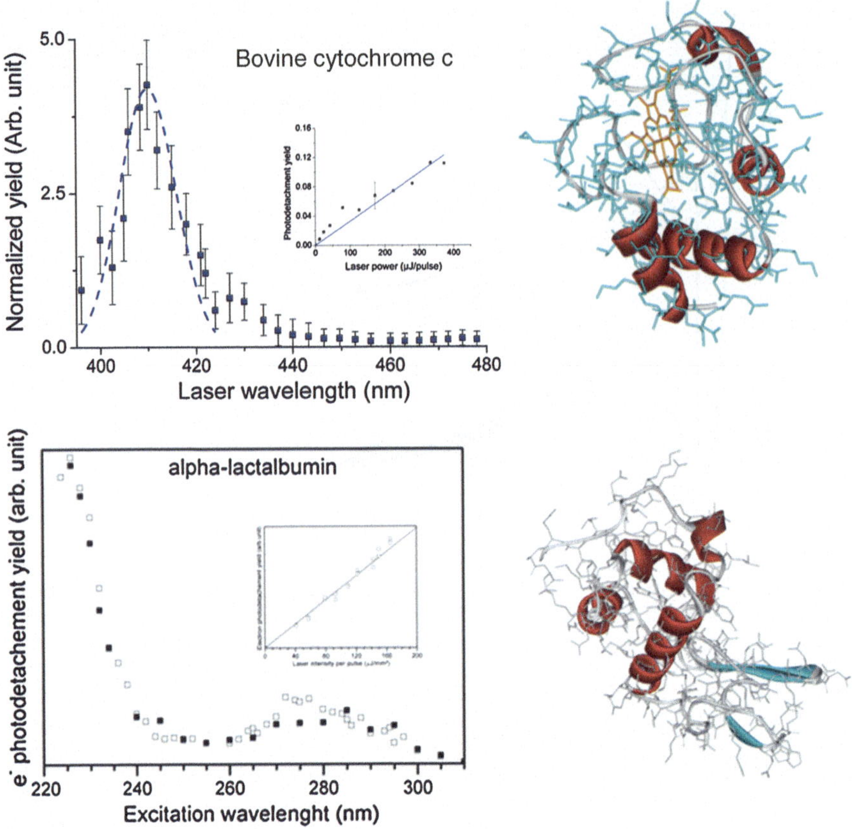

Fig. 5.4 Electron photodetachment yield as a function of the laser wavelength for bovine cytochrome c and alpha-lactalbumin proteins in the non-reduced form (*open squares*) and reduced form (*black squares*). *Insets*. Electron photodetachment yield as a function of laser intensity at 415 nm for cytochrome c (*top*) and 280 nm for alpha-lactalbumine (*bottom*). *Right panel* pdb X-ray structures of the proteins

visible or near-UV lasers for ion excitation and thus to extend photoexcitation to new classes of molecules without VUV sources.

The impact of oligosaccharide fluorophore tagging on the dissociation behavior following laser irradiation has been investigated with regard to fragmentation yield and dissociation pattern in negative EPD mode at 355 nm [4]. Grafting lacto-N-difucohexaose (LNDFH-II) with 7-amino 1,3-naphtalene-disulfonic acid (AGA) allows to ensure the formation of a doubly deprotonated precursor ion for subsequent preparation of an oxidized species by electron detachment under laser irradiation at 355 nm (Fig. 5.5). Some differences may be underlined when comparing the fragmentation pattern between the oxidized species and its singly deprotonated even electron congener. CID dissociation of the latter generates

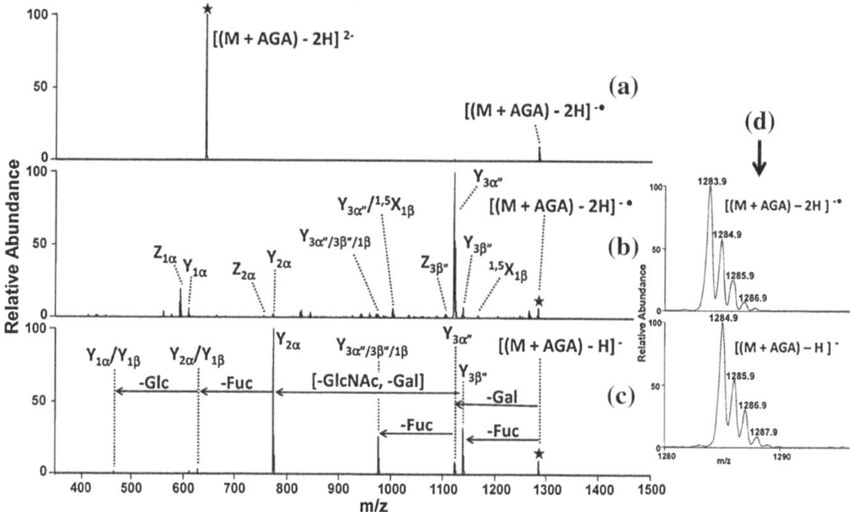

Fig. 5.5 EPD of LNDFH-II derivatized with AGA: (**a**) shows the EPD spectrum upon irradiation of doubly deprotonated LNDFH-II using 20 laser pulses resulting in the production of the $[M + AGA - 2H]^{-\bullet}$ radical species; (**b**) is the CID of the resulting radical ion; (**c**) the CID of $[LNDFH\text{-}II + AGA - H]^-$; and (**d**) which shows the expanded regions around the precursor ions show the 1 Da mass difference between the radical ion and conventional closed-shell deprotonated species. A star symbol (*) is used to signify the precursor ion. (Adapted with permission from Ref. [4]. Copyright 2008 American Chemical Society)

mainly y-type fragment ions, while the EPD spectrum reveals a few additional z- and cross-ring x-type cleavages.

5.3 UV Photodissociation of Protonated Peptides

5.3.1 Relaxation and Fragmentation of Electronic Excited Peptides: The Example of Leucine-Enkephalin

Leucine-enkephalin (Tyr Gly Gly Phe Leu) has been used in a number of earlier studies of peptide dissociation in the gas phase [5]. It contains two chromophores (phenyl and phenol groups) that absorb UV light. Figure 5.6 compares collision-induced dissociation and UV photodissociation (UVPD at 220 nm) spectra obtained for this protonated peptide. The CID spectrum (a) is dominated by a-, b- and y- fragments (cf. Fig. 5.6a). The spectrum obtained using $\lambda = 220$ nm (Fig 5.6b) is different. The spectrum shows a peak at *m/z* 449 (loss of 107 Da). This peak, which is not observed in the CID spectrum, corresponds to loss of the neutral tyrosine side chain (see Scheme 1). In addition to the losses of 107, loss of

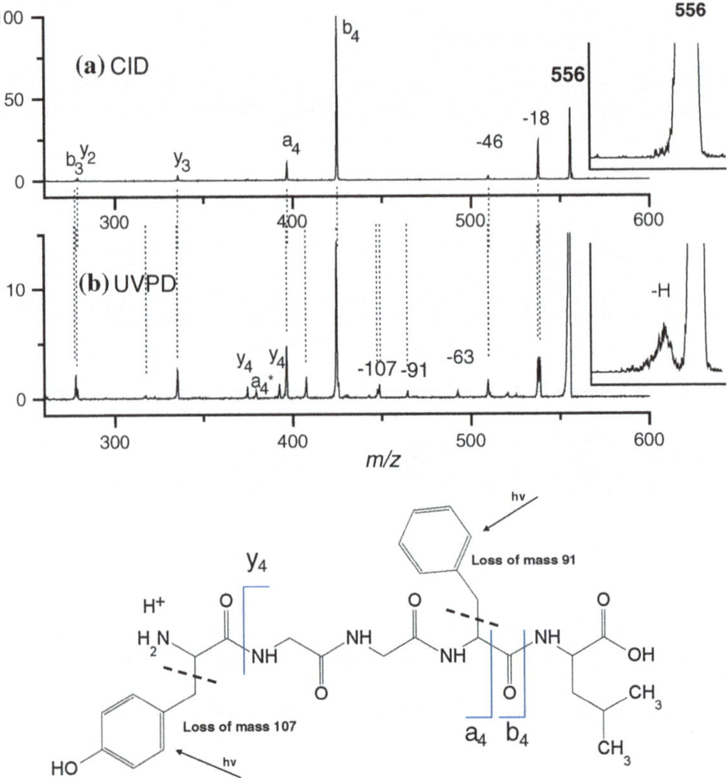

Fig. 5.6 CID and UVPD mass spectra of leucine-enkephalin (m/z 556 was isolated in the trap before dissociation), **a** CID, **b** UVPD at $\lambda = 220$ nm. The inserts in the right part of the figure show zooms of the parent peak region to display hydrogen losses (these spectra were obtained in the zoom scan mode of the trap). *Bottom* Structure of leucine-enkephalin with an indication of the main observed fragments

$$\textit{electrospray} \rightarrow \textit{closed shell ion} \xrightarrow{\textit{light excitation}} \textit{open shell ion} \xrightarrow{\textit{excitation}} \textit{radical fragmentation}$$

Scheme 1 Photoformation and activation of radical ions

91 Da, corresponding to cleavage of the phenylalanine side chain, is observed (peak at m/z 465). All the fragments (a, b, and y ions …) that were obtained in the CID [spectrum (a)] are detected in this spectrum (b). H-loss and the loss of 17 Da that may correspond to NH_3 loss are also only observed in UVPD.

The observation of specific dissociation channels (loss of 107 and 91 Da) that correspond to cleavage close to the chromophore is an indication of a fast dissociation occurring prior to IVR. Two processes can be responsible for the specific cleavage: (1) a direct dissociation in a dissociative electronic excited state. (2) electronic de-excitation of the ion, leading to the electronic ground state followed by a fast breaking of the $C\alpha$–$C\beta$ bond prior to IVR (see Fig. 5.6c). The loss of 107 and 91 Da lead to the formation of a radical cation whose subsequent fragmentation is discussed below.

The H-loss results from a laser-induced π–π^* transition in the chromophore and to a direct dissociation resulting from the coupling between the electronic excited π^* state and a dissociative σ^* state (as illustrated in the inset of Fig. 5.3 for phenol).

The last family of fragments observed in UVPD spectra corresponds to fragmentation of the peptide backbone and are similar to those observed in CID. These fragments are due to IC from the excited states to the ground state, followed by IVR and eventually fragmentation.

5.3.2 Peptide Sequencing with UVPD

For protein identification, peptides are typically generated by enzymatically digesting proteins. Peptides digests are then analyzed by coupling of chromatography and mass spectrometry. Most often, the products of peptide ion fragmentation are identified by comparison with predicted fragments derived from protein sequence databases. Searching for the most informative fragmentation patterns has led to the development of a vast array of activation modes that offer complementary ion reactivity and dissociation pathways, among which UVPD. Figure 5.6 has shown the apparition of specific product ions related to fragmentation processes faster than energy redistribution. Increasing the photon energy (for example using ArF and F_2 excimer VUV lasers) permits to enhance fragmentation with the opening of new fragmentation pathways. Figure 5.7 shows mass spectra of peptides fragmented using 157 nm photodissociation. Almost complete series of backbones fragment ions allowing peptide identification were obtained. Every spectrum is dominated by *x*, *v*, and *w* fragments, which are not observed in CID. About 157 nm allows excitation of peptides without aromatic acid, and it can also promote electrons to different excited states and ionized the peptides. These processes are followed by relaxation resulting in *x*, *v*, and *w* ion products which are not observed after CID and at excitation around 266 nm.

5.4 Photoinduced Radical Chemistry

5.4.1 Background

While CID fragmentation of biomolecular ions leads mainly to closed-shell fragment ions, fragmentation in electronic excited states may lead to homolytic cleavages (as observed for aromatic amino acid side chain loss in leucine-enkephaline). This photogenerated radical can be isolated in the mass spectrometer and further activated, leading to radical induced chemistry.

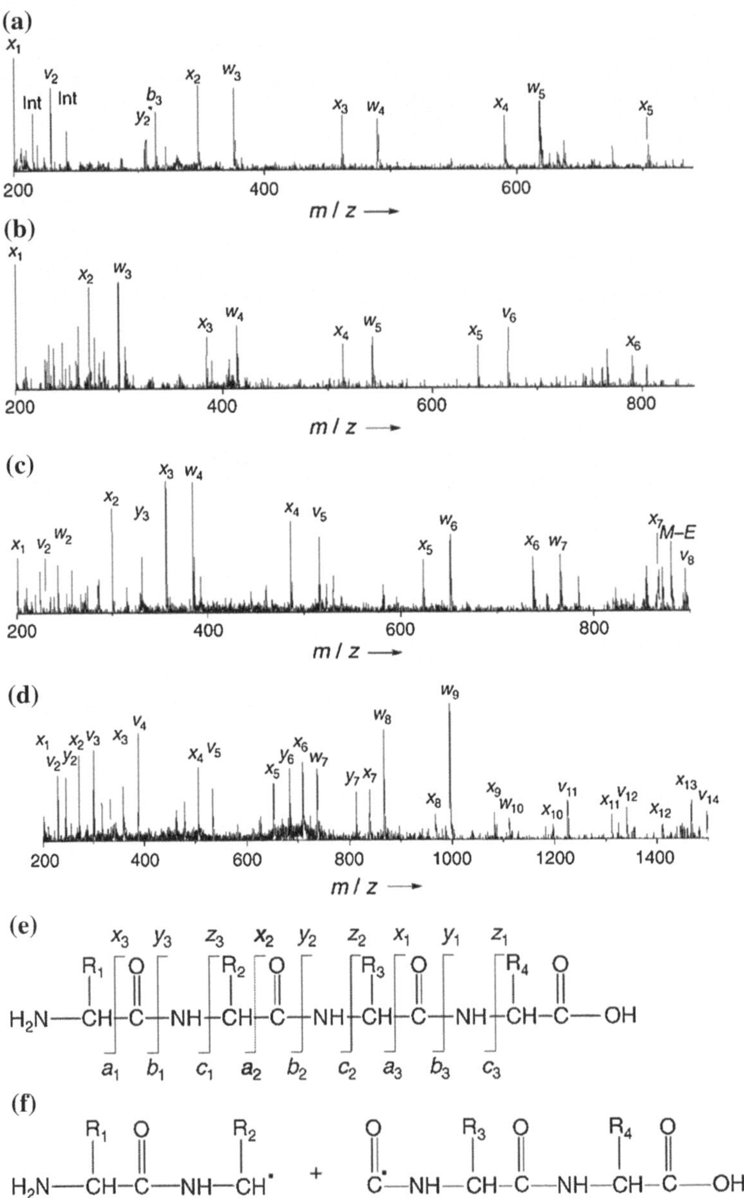

Fig. 5.7 MALDI 157 nm photodissociation tandem time-of-flight mass spectra of a series of peptides containing C-terminal arginine. **a** ALELFR; **b** LFEELAR; **c** IENHEGVR; **d** EG-VNDNEEGFFSAR. Peaks labeled Int are internal fragments. The (*) label represents the loss of one NH3 group. **e** Standard nomenclature of peptide fragmentation; **f** products of homolytic radical cleavage of a sample peptide. Adapted with permission from Reilly and coll. Angew. Chem. Int. Ed. 2004, 43, 4791–4794. Copyright © 2004 WILEY–VCH Verlag GmbH and Co. KGaA, Weinheim

Other schemes have been developed to produce radical ions including electron photodetachment from polyanions [6, 7], fragmentation of metal–biomolecular complexes [8, 9] and fragmentation after chemical derivatization [10, 11].

5.4.2 UV Photodissociation of Phosphopeptides

Phosphorylation is a ubiquitous process that plays a crucial role in most of the essential biological events, i.e., transcription, signal transduction, or apoptosis. When protonated precursor ions of phosphorylated peptides are selected, CID/MS^2 spectra exhibit intense ions originating from gas-phase elimination of phosphate ester (HPO_3, −80 Da) or phosphoric acid (H_3PO_4, −98 Da). While these characteristic neutral losses facilitate the discrimination of phosphorylated peptides, the localization on the peptide backbone of the amino acid residues bearing the phosphate moiety often fails as informative fragments ions are usually of low intensity, or lacking. Laser irradiation at 220 nm, then followed by a CID experiment, was investigated allowing the production and activation of radical ions.

The N-terminus acetylated and C-terminus amidated form of peptide Ac-VY-KpSPVVSGDTSPRHL-amide was studied. The following procedure was used to perform UVPD-CID MS^3 experiments [12]. Precursor ions are first isolated in the trap. After isolation, they are irradiated with the UV laser. After irradiation, the photofragment obtained after Tyr side chain elimination (loss of 107 Da) is isolated in the trap. These selected ions are fragmented by CID using helium as collision and damping gas. The ions are finally ejected from the trap, and the resulting mass spectrum is recorded.

In CID mode (Fig. 5.8a), the doubly protonated peptide initiates two major neutral losses, i.e., H_2O (m/z 923) and H_3PO_4 (m/z 883), as well as cleavages of the peptide backbone of the b and y-type series. Two doubly charged fragment peaks at m/z 918 and m/z 878, attributed, respectively, to the neutral losses of 28 and 107 Da, are the main species observed in the LID mass spectrum of the doubly protonated precursor ion (Fig. 5.8b). The first fragment corresponds to the elimination of a CO group. The second peak is due to the side chain cleavage of the tyrosine leading to the formation of the radical ionic moiety ($[MY^{\bullet}]^{2+}$).

The LID/CID-MS^3 fragment ions pattern of the $[MY^{\bullet}]^{2+}$ cation is shown in Fig. 5.8c. The direct loss of the phosphate group is fully inhibited. Moreover, no secondary loss of this moiety from the backbones cleavages, dominated by b and a species, is detectable. Satellite peaks corresponding to neutral losses of 16 and 44 Da from the two series of b and a fragment ions are systematically observed and account for the large density of peaks. They may originate from sequential losses of the methyl moiety and of the acetyl groups from the acetylated peptide N-terminus. Comparison of Fig. 5.8a and c outlines the different chemistry observed for closed and opened electron shell molecular ions.

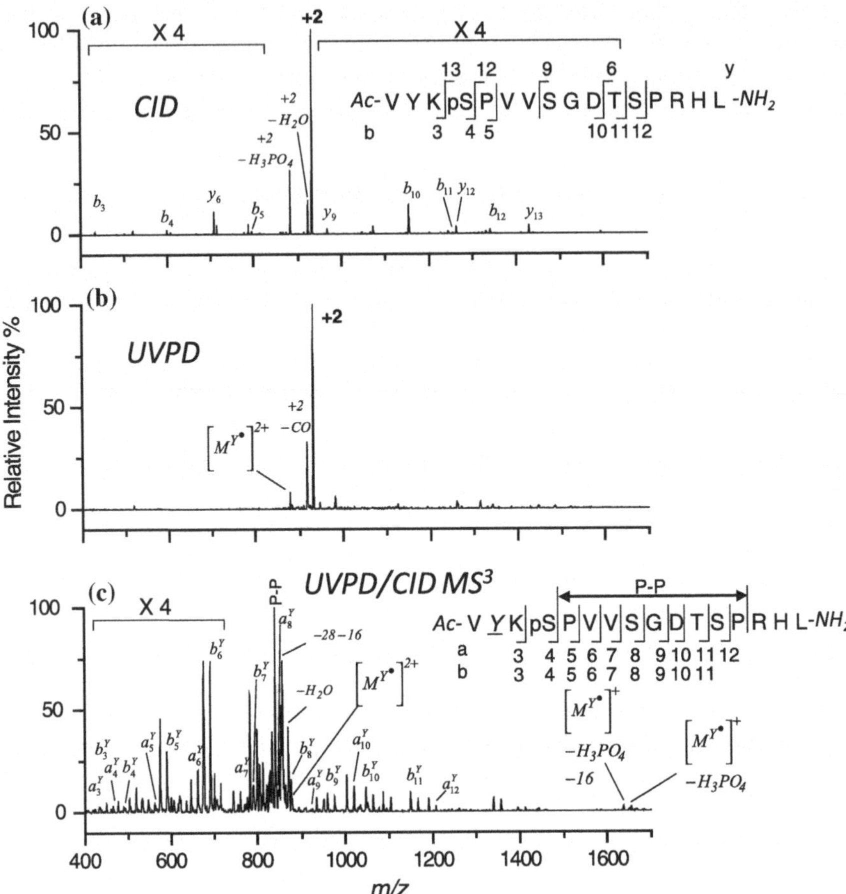

Fig. 5.8 **a** CID and **b** UVPD ($\lambda = 220$ nm) fragment ion spectra of doubly protonated phosphorylated peptide Ac-VYKpSPVVSGDTSPRHL-amide (m/z 932 was isolated in the trap before dissociation). **c** UVPD/CID-MS^3 spectrum of the $[M^{Y\bullet}]^{2+}$ fragment ion (m/z 878)

5.4.3 Activated-Electron Photodetachment Dissociation

Another approach to generate radical ions is to start with polyanions and to use photoactivation to induce electron loss. The oxidized radical ion is then isolated and further activated by collisions (activated-electron photodetachment dissociation, a-EPD) (Scheme 2).

The same phosphopeptide described in previous section is taken as example. The spectrum in Fig. 5.9a obtained after UV irradiation shows the formation of the $[M\text{-}2H]^{-\bullet}$ ion generated by laser-induced charge reduction from the $[M\text{-}2H]^{2-}$ precursor ion. The collisional activation spectrum of the $[M\text{-}2H]^{-\bullet}$ ion is presented in Fig. 5.9b and exhibits abundant a^- ions and x^- ions. A significant neutral

Scheme 2 Activated-electron photodetachment dissociation (a-EPD)

$$[M-2H]^{2-} + h\nu \rightarrow [M-2H]^{-\bullet} + e^{-}$$

$$[M-2H]^{-\bullet} \xrightarrow{\text{collision activation}} \textit{fragments}$$

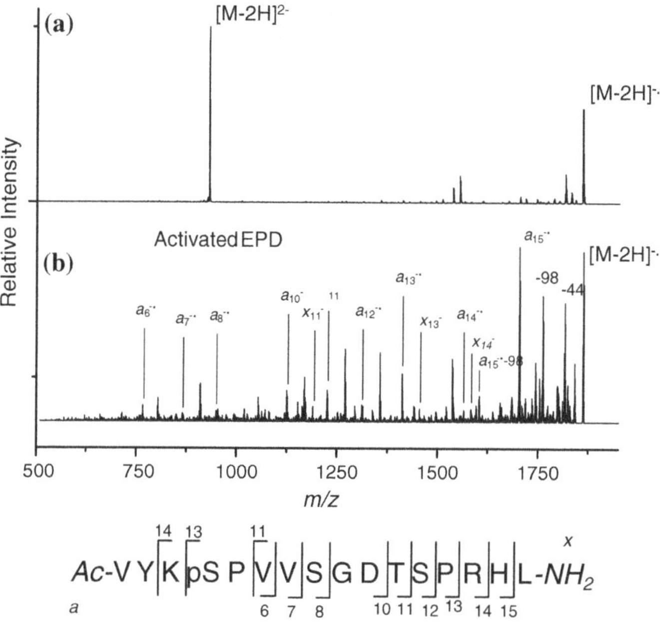

Fig. 5.9 **a** Mass spectrum obtained after laser irradiation ($\lambda = 262$ nm) of the doubly deprotonated phosphorylated peptide Ac-VYKpSPVVSGDTSPRHL-amide. **b** activated- EPD spectrum of the phosphopeptide. The $[M\text{-}2H]^{-\bullet}$ ion obtained by electron detachment from the $[M\text{-}2H]^{2-}$ ion was isolated and activated by collisions

loss (−98 Da) is also observed attributed to the elimination of phosphoric acid. It has to be pointed out that a_6^-, a_7^-, a_8^-, a_{10}^-, a_{11}^-, a_{12}^-, a_{13}^- a_{14}^- and all the three x^- fragments have retained the phosphate group. Consequently, the pattern of fragment ions allows one unambiguously to delineate this modification on the serine 4. a-EPD is a method particularly suitable for acidic molecules like DNAs and sugars and synthetic acidic polymers [13–16].

5.5 Coupling of UV-Visible Fragmentation with Separation Methods

5.5.1 Coupling of Photodissociation with Liquid Chromatography

The coupling between liquid chromatography and tandem MS (LC/MS/MS) has become the most popular technique for quantitative determination of chemical or biological compounds in complex matrices because of its superior specificity and sensitivity. The triple-quadrupole mass spectrometer operating in selected reaction monitoring (SRM) mode is particularly considered a core technology to develop high throughput quantitative assay of pharmaceuticals in blood or urine during pharmaco-kinetic investigations, or of contaminants in environmental matrices. In SRM mode, quadrupole 1 filters only the molecular ion of the target compound, which is then subjected to collision-induced dissociation, while quadrupole concomitantly monitors one or more product ions. The concomitant filtering of a given precursor ion and its specific product ion is called transition.

However, strong limitations remain regarding the quantification of highly diluted compounds in complex matrices like for instance blood biomarkers leaking from tumoral tissues. The likelihood of intense interfering signals within the targeted channels is in this case very high. Collision-induced dissociation (CID) is at that time the exclusive fragmentation mode available on commercial triple-quadrupole instrument. CID excitation is objectively efficient in opening numerous and complementary informative fragmentation channels. However, the close CID activation thresholds for co-eluting compounds that share similar precursor and product ion m/z ratio do not enable them to be discriminated. Photodissociation initiated through photon absorption is in contrast inherently much more specific since only the molecules absorbing at the appropriate wavelength will fragment. In the work described below, a laser beam at 532 nm has been focused in the middle quadrupole of a commercial triple-quadrupole instrument (see Fig. 5.10) [17].

The commercial QSY® 7 dye was used to label peptidic hormone oxytocin (cysteine-containing peptide). This dye presents a significant absorbance at 532 nm. The photodissociation spectrum at 532 nm of chromophore-tagged oxytocin is shown in Fig. 5.10b). Major product ions are issued from cleavages inside the chromophore that retained the fixed-cationic charge.

To compare the relative merits of photo-SRM and CID-SRM detection selectivity of peptides, chromophore-tagged oxytocin was spiked in a whole human serum trypsin digest to draw a calibration curve over a 0.41 to 20.5 μg per mL of serum. 1327.5/465.5 and 1327.5/360.8 transitions were selected to track chromophore-tagged peptide in CID and photo-SRM modes, respectively.

Figure 5.11 shows the reconstructed ion chromatograms recorded in CID-SRM (a) and in photo-SRM (b) for tagged oxytocin. In CID, a major interference is especially noticeable at 6.2 min. In contrast, no major interference is detected

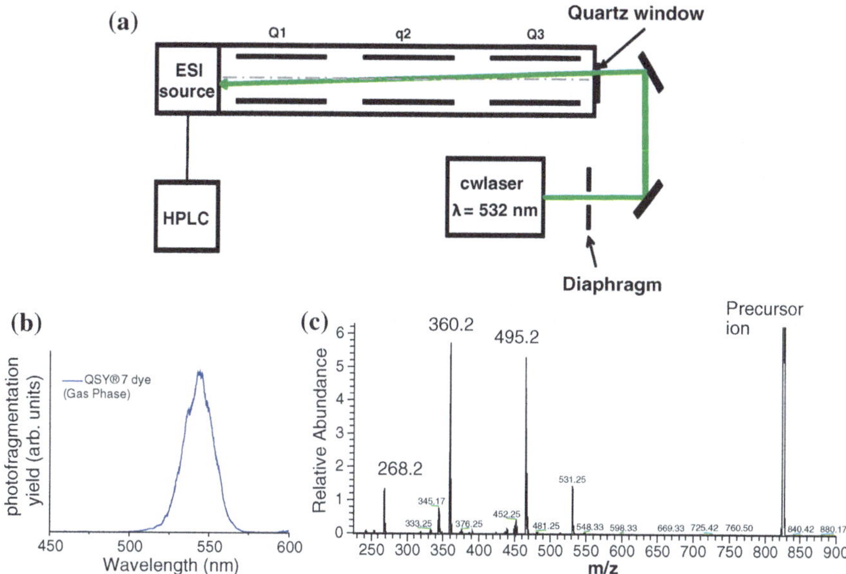

Fig. 5.10 **a** Schematic of the experimental setup. **b** Photofragmentation yield spectrum of the QSY® 7 dye in the visible range. **c** Photofragmentation ($\lambda = 532$ nm) MS^2 spectrum of $[M + H]^+$ ion

moving to photo-SRM mode that features the very high specificity of the 532-nm laser excitation process.

This investigation clearly demonstrates that both the selectivity and detection level may markedly be improved during SRM monitoring by replacing the classical mode of activation based on gas collision by a photon excitation providing a judicious overlap between the absorbing properties of the target molecule and the excitation wavelength of the laser beam.

5.5.2 Coupling of Photodissociation with Ion Mobility Spectrometry

Ion mobility is a gas-phase technique separating ions based on their collision cross-section with a neutral gas. Ions are accelerated by a DC field in a drift tube filled with a neutral gas. Compact ions (with small collision cross-sections) travel faster than more extended ions. An instrument that combines ion mobility spectrometry (IMS) separations with tandem MS (MS^n) is shown in Fig. 5.12. The instrument is made of a drift tube connected to a linear ion trap that has been interfaced to a pulsed F_2 excimer laser (157 nm).[19]I on gates positioned in the front and the back of the primary drift region allow for mobility selection of

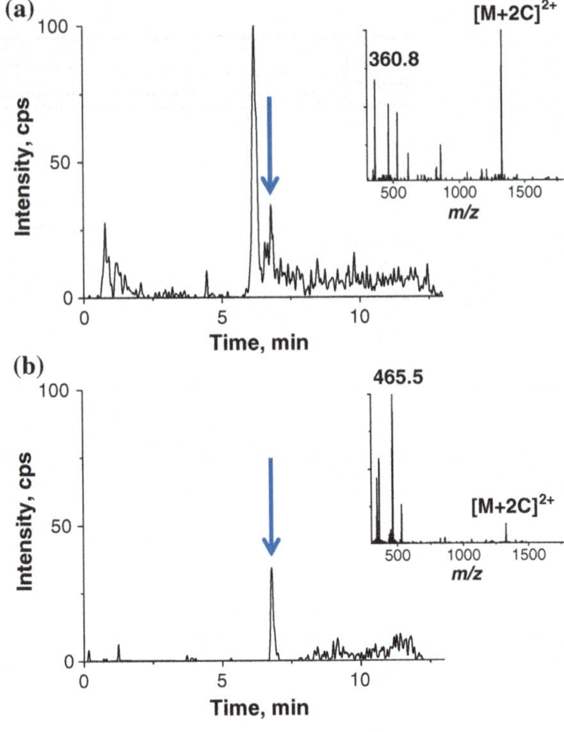

Fig. 5.11 LC-MS/MS chromatograms of human serum digest. **a** SRM analysis of the most intense SRM transition (1327.5/360.8) for the tagged oxytocin with fragmentation induced by collision (CID) and (**b**), SRM analysis of the most intense SRM transition (1327.5/465.5) for the tagged oxytocin with fragmentation induced by laser (visible photodissociation) $\lambda = 532$ nm. The *arrows* show the peaks corresponding to the molecule of interest, tr = 6.5 min. Inserts show MS^2 spectra of oxytocin with the corresponding method

specific ions prior to their storage in the ion trap, mass analysis, and fragmentation. Mobility separation of precursor ions provides a means of separating the isomers before photoexcitation.

Melezitose and raffinose are two isomeric trisaccharides differing only in sequence and linkage. The drift time distributions for the precursor ions are shown in Fig. 5.12b. The two peaks associated with the higher- and lower mobility ions centered at 16.2 and 16.7 ms are assigned to the $[M + Na]^+$ melezitose and raffinose ions, respectively.

Figure 5.12c shows the VUVPD spectra for $[M + Na]^+$ melezitose and raffinose ions. Because the isomers are composed of the same three monosaccharide units, many of the fragments are common to both species. These fragments can be attributed to fragmentation on one of the six-membered rings. Photodissociation of the higher mobility $[M + Na]^+$ melezitose ions results in four peaks not produced by fragmentation of raffinose. These fragments appear at m/z values of 317.1, 333.1, 421.1, and 453.1. The technology described here shows significant promise for analyzing complex mixtures of geometric isomers.

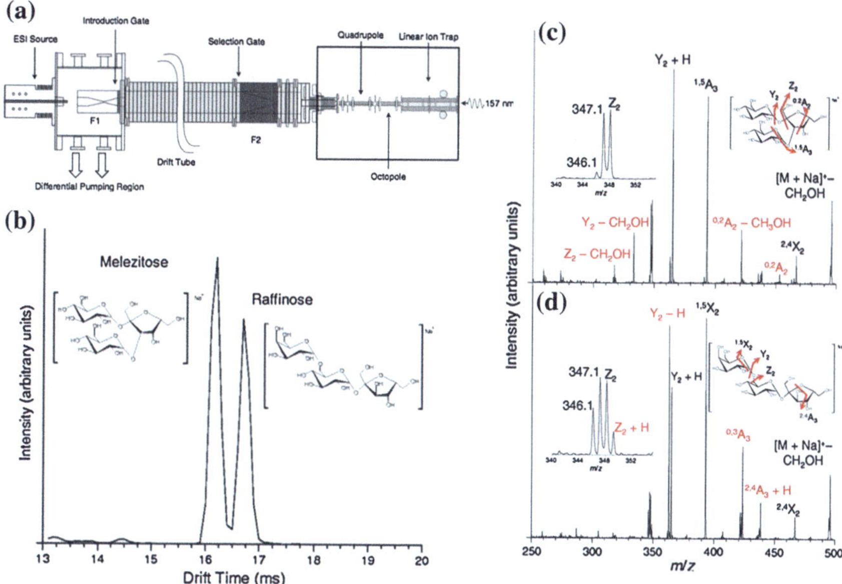

Fig. 5.12 **a** Schematic of the IMS-MS instrument. The ion source, ion gates, drift tube, and ion trap are labeled. As shown, the laser is aligned with the linear ion trap and can be used to introduce 157 nm light into the back of the ion trap. **b** The t_D distribution obtained for the analysis of the trisaccharide mixture. The two features in the distribution have been identified by comparisons to analyses of the pure compounds. Labels and schematic representations of the two different isomeric structures are provided for the identified features. VUV PD spectra of (**c**) melezitose and (**d**) raffinose selected from the mixture analysis at t_D values of 16.2 and 16.7 ms, respectively (panel b). Schematic representations of the two isomers displaying several fragmentation schemes are shown as insets. Separate insets show an expanded m/z region (340 to 355) to reveal differences in fragmentation patterns. Major fragment ions are assigned with diagnostic fragments shown in red. Adapted from [25], with kind permission from Springer Science and Business Media

5.6 Photodissociation of Biomolecular Ions with Ultrashort Pulses

Key parameters of femtosecond lasers are the following:

- the pulse duration (which is in some cases tunable in a certain range)
- the pulse repetition rate
- the pulse energy
- the wavelength (which is in some cases tunable in a certain range).

Bond-selective chemistry has been a goal of photochemists for decades. Ultrashort pulse duration permits ionization of peptides and ultrafast cleavage of chemical bonds at a rate faster than intramolecular energy redistribution [18], which opens the way to understanding and control relaxation pathways in tandem

MS. The high electromagnetic fields obtained with ultrashort pulses (femto- and atto-seconds pulses) result also in possible modification of the potential energy surfaces seen by electrons with a totally new physics not available for longer pulses.

5.6.1 Femtosecond Laser-Induced Ionization/Dissociation of Peptides

Beyond multiphoton dissociation and ionization easily achieved with the high-energy density obtained with ultrashort pulses, intense femtosecond pulses can cause ultrafast electron loss through a process known as tunneling ionization. The condition required to achieve tunneling ionization is that the electron must be able to acquire sufficient energy to overcome the binding energy within a single cycle of the optical field. This process is illustrated in Fig. 5.13. Upon ionization of a protonated peptide, the species formed is a distonic cation $[M + H]^{+} \rightarrow [M + H]^{2+\bullet}$, which is susceptible to both proton- and radical-directed fragmentation pathways. Fs excitation MS spectrum of the peptide GAILAGAILR which contains no aromatic residues is shown in the lower panel in Fig. 5.13. Nearly complete series of b-ions and a-ions are observed. Another notable feature in the fs-photodissociation spectrum is the presence of x and z ions, which are related to radical fragmentation and is generally observed in electron excitation methods (ECD, EDD) or EPD (see Sect. 5.4.3).

In conclusion, a femtosecond laser can activate any class of molecules via tunneling ionization, independent of their absorption spectra. Another benefit of fs-UVPD is that it creates a radical to access several non-ergodic dissociation pathways.

5.6.2 Femtosecond Pump-Probe Experiments on Trapped Flavin: Optical Control of Dissociation

With their short pulse duration, comparable with intramolecular dynamics, the combination of femtosecond pulses with MS is one of the most promising technology for laser control of chemical reactions. Femtosecond pump-probe experiments were performed on flavin biomolecules isolated in an ion trap (see Fig. 5.14). Mass spectra of the photoinduced fragments show that the ion product yields can be modified using two-color two-photon excitation. In particular, when an infrared probe pulse (810 nm) is added subsequent to the first excitation step (excitation at 405 nm of the S1 state of flavin mononucleotide), branching ratios between lumichrome and lumiflavin production are inverted relative to the single excitation case.

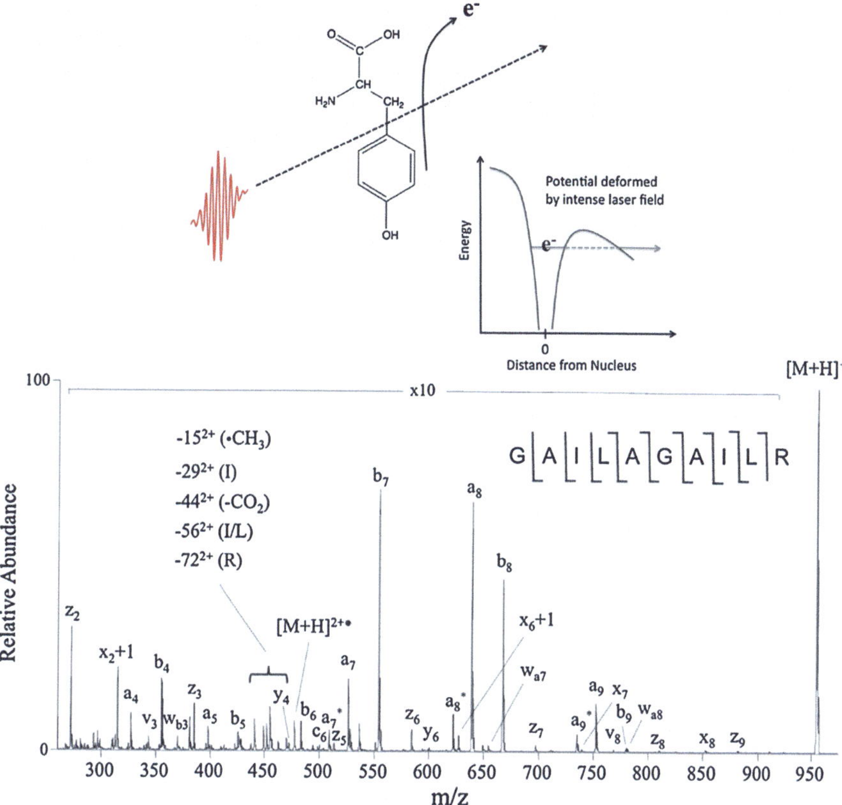

Fig. 5.13 (*Upper Panel*) As an ultrafast laser pulse passes by the analyte, the intense electric field deforms the potential felt by electrons within the molecule. As a result, the electron that is most polarizable is able to escape from the analyte, leaving behind a photoionized radical site. (*Lower Panel*) fs-LID MS/MS spectrum for GAILAGAILA. Adapted with permission from Ref. [26]. Copyright 2012 American Chemical Society

5.7 Future Directions

The development of MS techniques for analyzing biological samples, in particular tandem MS using collisional induced dissociation (CID), has played a vital role in establishing modern day proteomics. Analysis of the fragmentation pattern of a peptide to decipher the amino acid sequence is, for instance, useful for an a priori unknown protein identification or post-translational modification positioning. However, identification and structure analysis of high molecular weight analytes (proteins, DNA, polysaccharides ...) often diluted in complex mixtures is still a daily challenge in MS. By using alternative activation methods instead of conventional CID, a greater degree of backbone fragmentation can be achieved. UV-based activation methods are becoming increasingly well established and promote

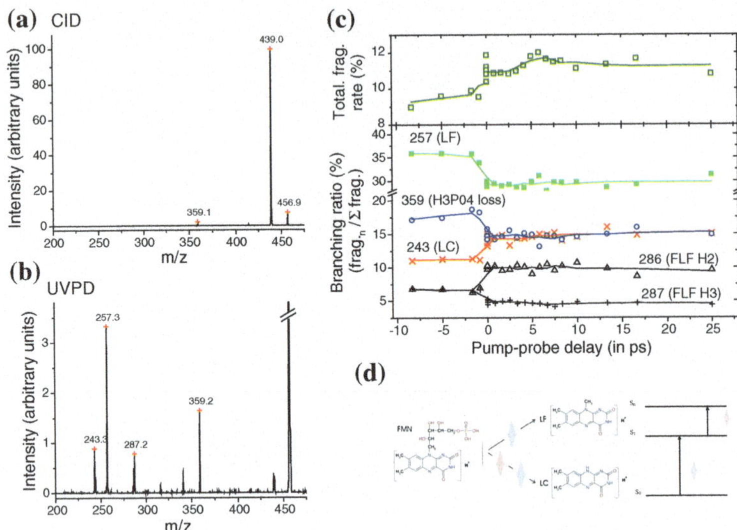

Fig. 5.14 Mass spectra of trapped protonated FMN (m/z 457). **a** Collision-induced dissociation (CID). **b** Ultraviolet photodissociation (UVPD) for a pump at $\lambda = 405$ nm. **c** Branching ratio measured for the main fragments of FMN as a function of the pump-probe delay. The total fragmentation rate $(\sum fragments/(parent + \sum fragments))$ is also plotted. *Solid lines* are smoothed curves of the experimental data. **d** When an infrared probe pulse (810 nm) is added subsequent to the first excitation step (excitation of the S_1 state of flavin mononucleotide, FMN, at 405 nm), branching ratios between lumichrome (LC) and lumiflavin (LF) production are inverted relative to the single excitation case. A pictoral diagram of energy levels of FMN is given to summarize the pump-probe experiment. Adapted with permission from Ref. [27] Copyright 2008, American Institute of Physics

greater diversity in the fragmentation pathways compared with those exhibited up on CID. A high flux of photons combined with a large absorption cross-section induces highly efficient and fast dissociation rates. Technological progress in photon sources have delivered a wide range of photon energy ranging from infrared (IR) to X-rays, tunable excitation duration from atto second pulses to continuous excitation, while the density of photons (source power) can easily be monitored and modified. Depending on the wavelength, the specificity of photon/ ion interactions can be used to finely control the location and amount of energy introduced in the molecular ion and eventually fragmentation patterns. We have seen in this chapter that visible and UV excitation may result in direct fragmentation of the precursor ion in electronic excited states leading to fragmentation patterns different from those observed with slow heating methods (CID, IRMPD). Light can also be used to form radical ions through electron detachment, ionization or homolitic cleavages. This leads to radical chemistry with specific dissociation processes. One can then predict promising developments for MS coupled with UV photon activation. However, sequencing, identification and quantification of biomolecules by MS most frequently relies upon the coupling of separation method followed by tandem mass spectrometry. Systematic inclusion of photoactivation in MS protocols requires photoactivation periods compatible with separation methods (i.e., liquid chromatography) and high-throughput mass analysis. First results published in this direction show that these approaches will be favored by the use of high repetition rate lasers or continuous wavelength lasers.

In parallel to the use of photofragmentation for sequencing, identification and quantification of biomolecules, UV and visible lasers can be used to create new reactive intermediates that can be stored and manipulated in ion traps. The combination of chemical reactions in ion traps and laser spectroscopy is a promising way of developing and controlling photochemistry in well-defined environments [19, 20].

As already mentioned, bond-selective chemistry has been a goal of photochemists for decades. Control of the fragmentation pathways and thus chemical compounds formation by photon activation opens new perspectives from the fundamental as well as analytical viewpoints. Such control using laser wavelengths was recently demonstrated for a tryptophan silver complex (see Fig. 5.15) [21]. The results demonstrate that a photoinduced charge transfer occurs from the indole ring to the silver atom, leading to efficient silver loss and open the possibility to induce this process by appropriate choice of laser wavelength and/or by molecular structure.

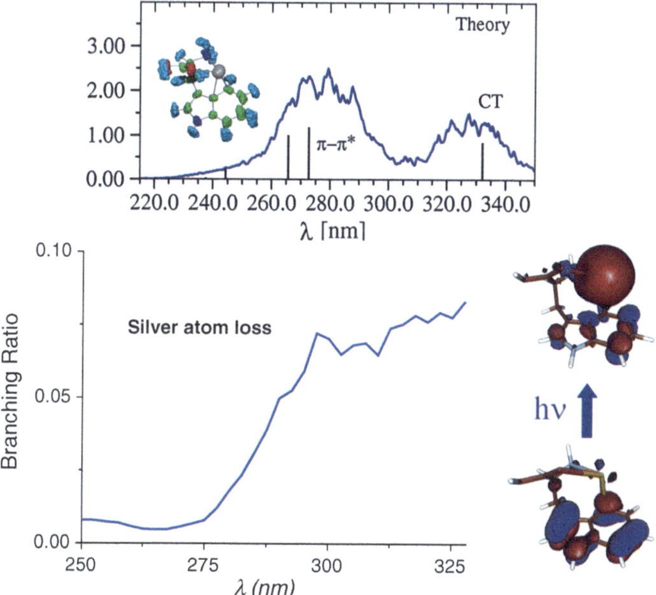

Fig. 5.15 (*Upper Panel*) Theoretical absorption spectrum for $[Trp\text{-}Ag]^+$. (*Lower Panel*) *left*. Branching ratios corresponding to the loss of the neutral silver. *Right*. The excitation responsible for the 332 nm band illustrates the charge transfer from the indole to the silver (*red* and *blue* colors label different signs of the lobes)

This example relies on the existence of a well-defined charge transfer type of excitations between the different moieties of the complex. The possibility that chemical reactions may be controlled by tailored femtosecond laser pulses has inspired many studies in molecular physics that take advantage of their short pulse duration, comparable to intramolecular dynamics. An example of pump-probe experiment was given in Sect. 5.6.2. A general control of the dynamics of fragmentation of biomolecular ions will certainly involve optimal control, using shaped femtosecond pulses to drive the process chosen. In Fig. 5.16, a femtosecond laser pulse shaper provides spectrally phase-modulated electric fields which are then iteratively optimized by a learning evolutionary algorithm. Direct

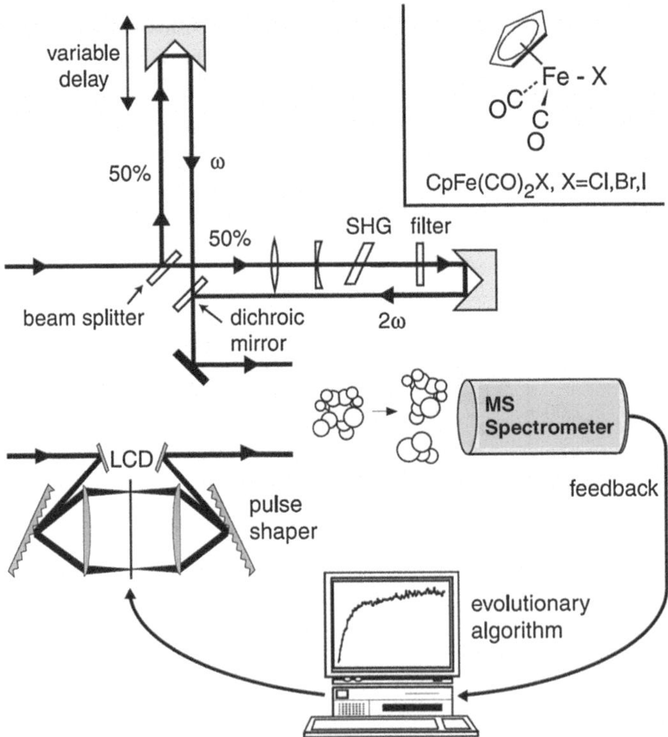

Fig. 5.16 Experimental setup. *Top part*-time resolved mass spectrometry is conducted by providing 400 nm pump and 800 nm probe laser pulses with a Mach-Zehuder interferometer. The probe arm can be delayed using a computer-controlled delay stage. After recombination, the pump and probe laser pulses are focused into the interaction region of a RETOF spectrometer. *Bottom part* for the automated quantum control experiment, a computer-controlled pulse shaper generates arbitrarily tailored femtosecond laser pulses. The experimental outcome (the photodissociation product yield) obtained with these pulses is used as feedback in an evolutionary algorithm. The electric field of the laser pulses is iteratively optimized until an optimal result is found. *Inset* The chemical structure of the molecules CpFe(CO)2X is indicated. The Cp ring (C5H5) is h5-connected to the central iron atom. Adapted from Ref. [28].Copyright 2001, with permission from Elsevier

experimental feedback from the photoproduct mass spectra is used in this search for specific optimization goals.

Another important aspect in UVPD is the dynamics of relaxation after photoexcitation. After photoexcitation, ionization and direct dissociation in the excited state compete with IC to the electronic ground state and with radiative de-excitation. In tandem MS experiments, only mechanisms leading eventually to fragmentation or charge modification are observed. Measurement of fluorescence after visible and UV excitation of trapped biomolecular ions was pioneered by Parks and coworkers [22]. Recording of fluorescence spectra (emission or excitation spectra), while only reported for molecules with high quantum yield yet, is an alternative to action spectra obtained by recording fragmentation yields. Relation between fragmentation yields and absorption spectra is an open (and unsolved) question inherent to all action spectroscopy. Recording absorption spectra for gas-phase molecular ions is a challenge due to the low density of ion clouds in ion traps (or other mass spectrometer). This was successively achieved for small metal clusters (see Fig. 5.17). Cluster ions stored in an ion trap interact with photons trapped in a cavity (cavity ring-down spectroscopy). The storage lifetime of photons in the cavity provides "direct" observation of extinction of light. Measurement of ultraviolet absorption of Ag_9^+ is shown in Fig. 5.17b).

UV/Visible excitation and photodissociation of biomolecular ions depends directly on the electronic structure and absorption of chromophores. DNA and proteins present natural chromophores (DNA bases and aromatic amino acids) that absorbs in the near-UV region. Some proteins are not transparent to visible light which is due to the presence of prosthetic groups (i.e., cytochrome c). For lipids and polysaccharides, absorption in the visible and near-UV regions is dependent on derivatization naturally occurring or added on reactive groups of the molecule. Grafting of chromophores allows to adjust optical properties of a given ion to specific requirements. This strategy was used to develop photo-SRM quantification method (Sect. 5.5.1). Another possibility to excite biomolecular ions without natural chromophore is to use high photon energy allowing excitation of electrons which present weak or no delocalization. In Sect. 5.3.2, we showed examples of MS spectra at 157 nm. This was extended to deep VUV using synchrotron radiations [23]. Bands due to p and σ electron excitations are observed. Higher energies allowing the excitation of inner shell electrons have also recently been explored. C, N, and O near-edge ion yield spectroscopy of 8 + selected electrosprayed cations of cytochrome c protein (12 kDa) was performed by coupling a linear quadrupole ion trap with a soft X-ray beamline (see Fig. 5.18). The photoactivation tandem mass spectra were recorded as a function of the photon energy. The partial ion yields extracted from recorded mass spectra show significantly different behaviors for single and double ionization channels, which can be qualitatively explained by different Auger decay mechanisms.

The previous chapter shows how infra-red spectroscopy can be used to decipher geometric structure of small peptides. The sensitivity of optical spectra to biomolecular conformation and the possibility of using gas-phase UV-Vis spectroscopy in structural biology are delicate to assess. A global strategy for the

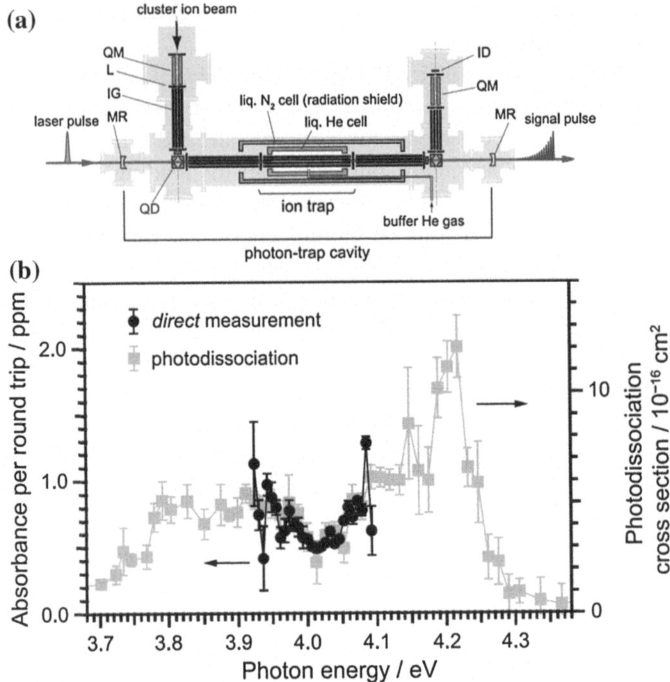

Fig. 5.17 **a** Experimental setup for photon-trap spectroscopy of mass-selected trapped ions. *QMs* quadrupole mass filters, *Ls* ion lenses, *IGs* octopole ion guides, *QDs* quadrupole deflectors, *ID* an ion-current detector *MRs* cavity mirrors. **b** Absorption spectra of Ag_9^+; direct measurement by photon-trap spectroscopy and measurement of an action spectrum of photodissociation. Both spectra were measured by an ion trap at room temperature. Adapted from Ref. [29]. Copyright 2009, with kind permission from Springer Science + Business Media

geometric characterization of biomolecules and their complexes should involve different experimental approaches as well as theory. Action spectroscopy could be coupled to IMS extending Clemmer and Reilly results to gas-phase spectroscopy on isomer resolved ions. Circular dichroism in ion yield has also promising potentials for analysis of chiral biomolecules. Recording fluorescence spectra opened the ways to FRET measurements which probe the distance between two chromophores attached to a molecule.

Finally, the trapping and characterization of optical spectra of biomolecular ions are initial steps toward studying biochemical processes in the gas phase. For example, charge transfer mechanisms play a significant role in mediating biological processes. It is a dynamical process that could be explored by two-color pump-probe spectroscopic experiments. Bio-photochemistry in the gas phase is an exciting future for spectroscopy.

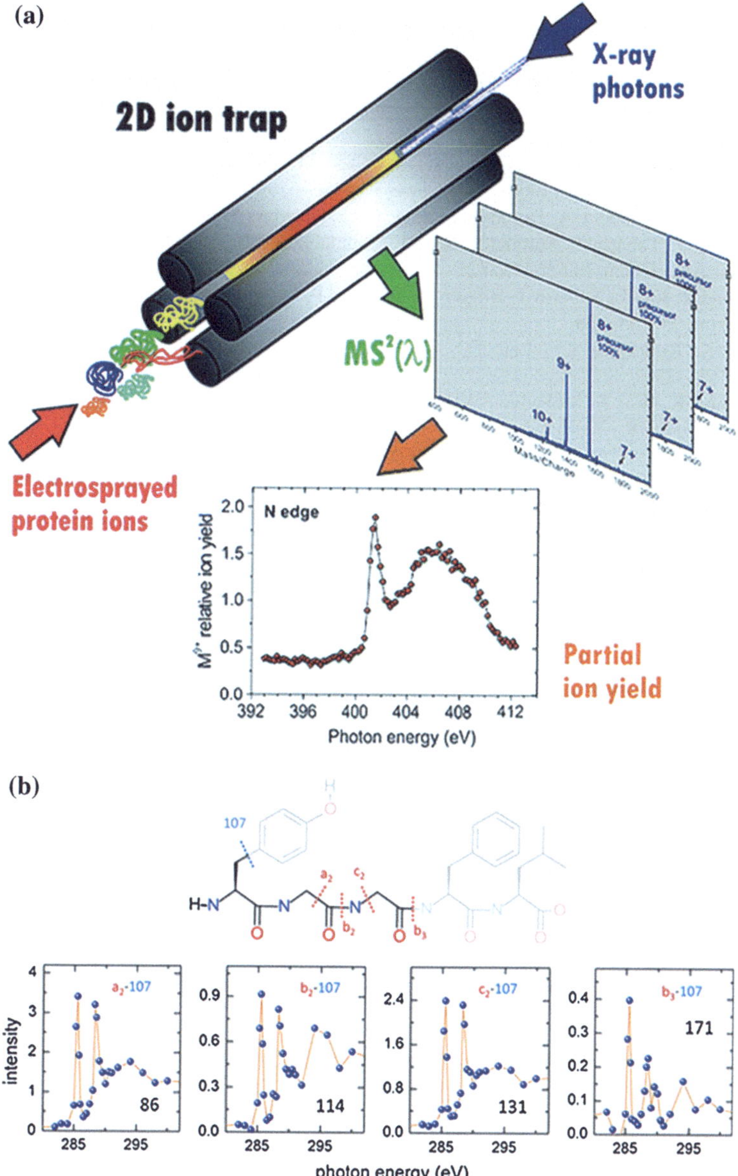

Fig. 5.18 **a** Principle of the experimental method for gas-phase protein inner-shell spectroscopy by coupling an ion trap with a soft X-ray beamline. Single (M^{9+}, 9+ charge state, m/z 1373–1375) C, N, and O K-edge photoionization yields of the 8+ charge state precursor (M^{8+}) of equine cytochrome c protein. **b** C K-edge Near-edge X-ray absorption fine structure spectra of [YGGFL+H]+ for fragments stemming from backbone scission followed by Y-side chain loss. Adapted with permissions from Refs. [30] and [31]. Copyrights 2012 American Chemical Society

References

1. Antoine R, Dugourd P (2011) Phys Chem Chem Phys 13:16494–16509
2. Creighton TE (1992) Proteins: structures and molecular properties. W. H. Freeman, San Francisco
3. Rizzo TR, Park YD, Peteanu L, Levy DH (1985) J Chem Phys 83:4819–4820
4. Wilson JJ, Brodbelt JS (2008) Anal Chem 80:5186–5196
5. Sztáray J, Memboeuf A, Drahos L, Vékey K (2010) Mass Spectrom Rev 30:298–320
6. Gabelica V, Tabarin T, Antoine R, Rosu F, Compagnon I, Broyer M, De Pauw E, Dugourd P (2006) Anal Chem 78:6564–6572
7. Antoine R, Joly L, Tabarin T, Broyer M, Dugourd P, Lemoine J (2007) Rapid Commun Mass Spectrom 21:265–268
8. Chu IK, Rodriquez CF, Lau TC, Hopkinson AC, Siu KWM (2000) J Phys Chem B 104:3393–3397
9. Barlow CK, Wee S, McFadyen WD, O'Hair RAJ (2004) Dalton Trans 2004:3199
10. Ly T, Julian RR (2007) J Am Chem Soc 130:351–358
11. Hodyss R, Cox HA, Beauchamp JL (2005) J Am Chem Soc 127:12436–12437
12. Lemoine J, Tabarin T, Antoine R, Broyer M, Dugourd P (2006) Rapid Commun Mass Spectrom 20:507–511
13. Girod M, Brunet C, Antoine R, Lemoine J, Dugourd P, Charles L (2012) J Am Soc Mass Spectrom 23:7–11
14. Racaud A, Antoine R, Dugourd P, Lemoine J (2010) J Am Soc Mass Spectrom 21:2077–2084
15. Larraillet V, Antoine R, Dugourd P, Lemoine J (2009) Anal Chem 81:8410–8416
16. Gabelica V, Tabarin T, Antoine R, Rosu F, Compagnon I, Broyer M, De Pauw E, Dugourd P (2006) Anal Chem 78:6564–6572
17. Enjalbert Q, Simon R, Salvador A, Antoine R, Redon S, Menaf Ayhan M, Darbour F, Chambert S, Bretonnière Y, Dugourd P, Lemoine J (2011) Rapid Commun Mass Spectrom 25:3375–3381
18. Kalcic CL, Gunaratne TC, Jones AD, Dantus M, Reid GE (2009) J Am Chem Soc 131:940
19. Rohr MIS, Petersen J, Brunet C, Antoine R, Broyer M, Dugourd P, Bonacic'-Koutecky V, O'Hair RAJ, Mitric R (2012) J Phys Chem Lett 3:1197–1201
20. Bellina B, Compagnon I, Houver S, Maitre P, Allouche A-R, Antoine R, Dugourd P (2011) Angew Chem Int Ed 50:11430–11432
21. Antoine R, Tabarin T, Broyer M, Dugourd P, Mitric R, Bonacic-Koutecky V (2006) Chem Phys Chem 7:524–528
22. Iavarone AT, Duft D, Parks JH (2006) J Phys Chem A 110:12714–12727
23. Bari S, Gonzalez-Magaña O, Reitsma G, Werner J, Schippers S, Hoekstra R, Schlathölter T (2011) J Chem Phys 134:024314
24. Sobolewski AL, Domcke W, Dedonder-Lardeux C, Jouvet C (2002) Phys Chem Chem Phys 4:1093–1100
25. Zucker S, Lee S, Webber N, Valentine S, Reilly J, Clemmer D (2011) J Am Soc Mass Spectrom 22:1477–1485
26. Kalcic C, Reid G, Lozovoy VV, Dantus M (2012) J Phys Chem A 116:2764
27. Guyon L, Tabarin T, Thuillier B, Antoine R, Broyer M, Boutou V, Wolf J-P, Dugourd P (2008) J Chem Phys 128:075103
28. Brixner T, Kiefer B, Gerber G (2001) Chem Phys 267:241–246
29. Terasaki A, Majima T, Kasai C, Kondow T (2009) Eur Phys J D 52:43
30. Milosavljevic AR, Canon F, Nicolas C, Miron C, Nahon L, Giuliani A (2012) J Phys Chem Lett 3:1191–1196
31. Gonzalez-Magana O, Reitsma G, Tiemens M, Boschman L, Hoekstra R, Schlathölter T (2012) J Phys Chem A 116:10745–10751

Index

A
AgGaSe2, 39, 82
Alpha–helical, 84, 86
Amide I, 73
Amide II, 73
Anharmonicity constant, 5
Anharmonic oscillator, 5
Anomericity, 75
Apex, 59
ArF, 21
Average power, 30

B
Beta–barium borate, 39
Biomolecules, 108, 111, 114
Birefringent material, 30
Blackbody radiation, 66, 67
Buffer gas, 58

C
Cavity, 25
CO2 lasers, 21, 72
Cold trap, 66
Consequence, 13
Consequence spectroscopy, 13, 16
Continuous wave, 25
Crown ethers, 78
Cryogenic, 82
Cryogenic traps, 66, 82
Cyclotron frequency, 51

D
Damage threshold, 38, 39
Degrees of freedom, 6
Dehmelt pseudopotential well, 62
Detection plates, 53
Diffraction grating, 30
Dipolar excitation, 53
Dissociative state, 17
Doping, 42
Dye lasers, 21

E
Electromagnetic radiation, 1
Electromagnetic spectrum, 2
Electron detachment, 94–96, 110
Electrospray ionization, 73
Endcap electrode, 61
Equipotential lines, 55
Etalon, 30
Excimer lasers, 21
Exciplexes, 34
Excitation plates, 53
External cavity, 46

F
Fluorescence, 17
Force constant, 3
Fourier transform ion cyclotron resonance, 50
Fragmentation, 93, 94, 96, 97, 99, 101, 104, 106, 108, 109, 110, 113
Franck–Condon, 86
Franck–Condon factor, 11
Franck–Condon principle, 9
Frequency, 1
FT–ICR, 50

G
Gain medium, 25, 26, 30, 31, 33, 35
Gas discharge lasers, 21

N. C. Polfer and P. Dugourd (eds.), *Laser Photodissociation and Spectroscopy of Mass-separated Biomolecular Ions*, Lecture Notes in Chemistry 83,
DOI: 10.1007/978-3-319-01252-0,

Glucose, 75
Glycosidic, 75
Glycosylated, 74
Gross selection rule, 4

H
Harmonic oscillator, 3
Heterostructure, 45
Hexapole, 65
HOMO, 9
Hydrogen–stretching, 78

I
Image current, 53
Indole, 80
Infrared active, 4
Infrared inactive, 4
Infrared multiple photon dissociation, 72
Internal conversion, 17, 85
Internal vibrational redistribution, 15
Ion activation, 101
Ion cyclotron frequency, 51
Ion dip spectroscopy, 18
IR predissociation, 82
IRMPD, 73
IR–UV ion dip, 82
Isotope labeling, 8, 86
Isotope substitutions, 84

K
Kinetic shift, 24
KrF, 21

L
3–level system, 27
4–level system, 27
Laplace equation, 54
Laser, 24, 95, 96, 101, 104, 108, 111, 112
Laser tomography, 58
Lithium niobate, 39
Littrow configuration, 46
Lorentz force, 50
Losses, 28
LUMO, 9

M
Macromotion, 57
Magnetron frequency, 52
Mass–selective transmission, 59
Mass spectrometry, 94, 99
Mass–to–charge ratio, 13
Mathieu equation, 56, 61
Metabolites, 88
Micromotion, 58
Morse potential, 5
MS–MS, 104
Multiplexing, 80
Multipoles, 65

N
Nd: YAG, 21
Nonlinear coefficient, 39, 40
Nonlinear optics, 21
Normal mode, 6

O
Octopole, 65
Operating line, 59
OPO, 21, 78
Orientation–patterned GaAs, 40
Origin band, 85
Output, 28
Output power, 30

P
Paul trap, 49, 54
Peak power, 30
Penning trap, 49, 50
Peptide, 73, 78
Phase matching, 39
Phosphate, 80
Phosphorylated, 74, 78
Photoexcitation, 96, 106, 113
Photon flux, 26
Population inversion, 26, 29, 32, 41
Potassium dihydrogen phosphate, 39
Potassium titanyl phosphate, 39
Prism, 30
Proteins, 94, 95, 99, 113
Pseudopotential, 62
Pulsed mode, 28
Pulse energy, 30
Pulse width, 30
Pumping source, 25

Q
Q–switch, 29
Quadrupolar potential, 52, 55
Quadrupole Ion trap, 55, 60

Quantization, 4
Quantum cascade lasers, 21
Quasi–phase matching, 39

R
Radiant power, 30
Radical ions, 101, 102, 110
Radiofrequency, 54
Reduced cyclotron frequency, 52
Reduced mass, 4
Repetition rate, 30
Resonant ejection, 64
RF–only mode, 59
Ring electrode, 61
Ruby laser, 35

S
Saccharides, 75, 81
Scan line, 59
Secular frequency, 64
Secular oscillation, 57
Simple harmonic motion, 56
Solid–state lasers, 21
Space charge, 58
Specific selection rule, 5
Spectral bandwidth, 30
Spectroscopy, 13
Spontaneous emission, 24
Stability limit, 63
Stimulated absorption, 24
Stimulated emission, 24

T
Ti–Sapphire, 21
Transition dipole moment, 4

U
UV chromophore, 85

V
Vertical transition, 9
Vibrational fundamental, 6
Vibrational quantum number, 4
Vibronic bands, 85
Vibronic transition, 16

W
Wavelength, 1
Wavenumbers, 1

Z
Zero–point energy, 4

The manufacturer's authorised representative in the EU is Springer Nature Customer Service Centre GmbH, Europaplatz 3, 69115 Heidelberg, Germany. If you have any concerns regarding our products, please contact ProductSafety@springernature.com

Printed and bound by CPI Group (UK) Ltd, Croydon, CR0 4YY
18/07/2026
02171007-0002